backyardchickens.com

W9-DEG-531

Beginner's Guide to Chickens

A Touch of the Good Life

LEE FABER

METRO BOOKS

NEW YORK

© 2009, 2010 (revised) Anness Publishing Ltd

This 2010 edition published by Metro Books,
by arrangement with Anness Publishing Ltd

Produced for Anness Publishing Ltd by
Omnipress Limited, UK
Cover design by Omnipress Limited, UK

Metro Books
122 Fifth Avenue
New York, NY 10011

ISBN: 978-1-4351-1767-9

Printed and bound in the United Arab Emirates

1 3 5 7 9 10 8 6 4 2

THE AUTHOR

LEE FABER is a native-born American who moved to
the UK in 1981. She has lived and worked in New York,
Florida, and London, and now resides in Wiltshire,
England. During her career she has been involved in
book editing and writing with an emphasis on health,
food and cooking. She has specialized in
Americanizing/Anglicizing books on a variety of
subjects for both US and UK publishers. She is also the
author of *Healthy Oils, Aloe Vera, Berries* and *Juices
and Smoothies* in this series. Lee is an accomplished
cook and has created many recipes.

NOTE

Although the advice and information in this book
are believed to be accurate and true at the time
of going to press, neither the authors nor the
copyright holder can accept any legal responsibility
or liability for any errors or omissions that may be
made nor for any inaccuracies nor for any harm or
injury that comes about from following instructions
or advice in this book.

CONTENTS

INTRODUCTION

Many of us dream of abandoning the rat race of the city and migrating to "the country" for a more satisfying self-sustaining existence. For some, our idyllic new life includes the desire to have a few chickens pecking around the backyard, providing us with the freshest imaginable eggs at little or no cost. However, this book isn't only for country dwellers. A backyard (or an accessible, secure flat roof) and a lot of determination can make this dream a reality, even in the inner city.

The thought of being more self-sufficient and having control over the food we eat can be a powerful motivator, especially with all the food scares (such as salmonella) we have gone through in recent years. Another strong motive is the desire to preserve endangered breeds of poultry.

As is the case with other pet animals, you really must have a fair amount of commitment before you buy chickens or agree to take on those offered as gifts. You have to be prepared to go out in all kinds of weather to feed and water them. You should find out in advance if anyone else in your family would be willing to take on part of the burden, and you have to make arrangements for your chickens when you go on holiday—you can't just leave them to forage for themselves.

It is easier to make a plan if you can decide what your primary motive for keeping chickens is—for example, having them only as pets, raising them for egg production and/or for the table, or breeding them to show.

Whatever your reason, whether it's to get back to basics, enjoy a new hobby or put healthy food on your table, this book will not only help you get started, but be a useful guide once your chickens are part of your life.

Before you embark on this very satisfying pastime, you should check carefully and make certain that there are no laws or regulations preventing you from keeping chickens, and if you also plan to have a cockerel or two, you might want to have a word with your neighbors and see how they react. Being awakened at 4 a.m. in the middle of the summer might not make you very popular!

A LITTLE HISTORY OF POULTRY

WHICH CAME FIRST—THE CHICKEN OR THE EGG?

The answer to the age-old question "Which came first, the chicken or the egg?" is neither. It was the dinosaur.

Despite its appearance as a ferocious predator, researchers have discovered that *T. Rex* was just an overgrown chicken.

A recent study has shown that the chicken is a direct descendant of the *Tyrannosaurus Rex* dinosaur. This new research follows a breakthrough study last year in which scientists reported the recovery and partial molecular sequencing of *T. Rex* and mastodon proteins.

Similarities between bone structure and the discovery of feathery remains on dinosaur fossils have been discussed previously, but scientists have now found the first molecular link. Both dinosaur studies examined samples of collagen, the main protein component of bone. Analysis of collagen has provided molecular evidence to support the theory that dinosaurs evolved into birds.

Originally it was thought that alligators were closely related to dinosaurs, but they are apparently more distant than ostriches and chickens and further back in the family tree.

The current study, carried out by scientists at Harvard University and reported in the journal, *Science* (April 27, 2008) claims that "analysis of a shred of 68-million year old *Tyrannosaurus Rex* protein—along with the proteins of 21 modern species—confirms that dinosaurs share common ancestry with chickens, ostriches, and, to a lesser extent, alligators."

POULTRY THROUGHOUT THE AGES

It is estimated that 11 billion chickens populate the planet. It would be impossible to count them all—from the few birds scratching a living outside a smallholder's dwelling to the thousands bred in huge industrial complexes. They are all, however, derived from the Red Jungle Fowl *Gallus gallus*, of southern and southeastern Asia.

DNA studies have shown that domestication of the Red Jungle Fowl probably began about ten thousand years ago. Poultry has historically been kept for several reasons: as clocks, for cock-fighting and for religious purposes, but only in the last 200 years or so, have they been selectively bred for eggs, meat, or exhibition.

It was probably the enterprising Egyptians who started the mass production of chickens and eggs for food. At the time of the Roman Empire, Europeans began to breed chickens for meat and eggs, but it wasn't until the mid-19th century that breeding for exhibition really became popular. Queen Victoria herself became interested in chickens, which popularized poultry-keeping as a hobby, and several color variants of British breeds were produced as a tribute to her.

Poultry-keeping was revolutionized after World War II. In 1945 less than 1 percent of laying hens in the UK were caged, but by 1986, 93 percent of the national flock was kept in cages. Battery cages in the UK and the rest of the European Union, but not the USA, are now being phased out, as more consumers have become opposed to the welfare implications. Free-range eggs are becoming more popular, and now nearly one in four British hens is free-range. Intensification for meat production has become equally extreme. Pre-war it took 126 days to produce a $4\frac{1}{2}$-pound bird; it now takes 42 days. There are also welfare concerns about broiler chickens and many people prefer to buy birds that have been reared in less intensive conditions.

Many people have decided that quality, welfare, and traceability are of the utmost importance, and prefer either to buy their meat from places such as local butchers, farmers' markets or to produce their own. Hens can make very good pets and have the added advantage that they produce eggs. Most people with a backyard will have enough room for a few hens, and there is nothing more satisfying than collecting home-produced eggs from happy and well-kept hens. With the addition of a cockerel, many people get added satisfaction from breeding their own poultry and conserving our valuable rare breeds.

According to some estimates, half a million households now keep chickens. Many of them are so proud of their flocks that they have taken to organizing "hen parties" in which current and prospective chicken owners gather in someone's backyard to cuddle the chickens and learn about how to keep them properly.

CHICKENS IN YOUR BACKYARD

CHICKEN BREEDS

There are many, many breeds and varieties of chickens. There is no way I can talk about all of them in a book this size, but they may be grouped into four basic classes: egg breeds, meat breeds, dual-purpose breeds, and fancy breeds.

EGG PRODUCTION BREEDS

The most noteworthy characteristics of the egg breeds are: small size, active and nervous temperament, early maturity, non-broodiness, good foraging habits, and sensitivity to cold.

Some representatives of this class are Leghorn, Minorca, Spanish, Andalusian, and Hamburg. All belong to the Mediterranean class, except the Hamburg, which is classed as European.

MEAT BREEDS

Large size, gentle disposition, slowness in movement, poor foraging proclivities, as a rule poor laying qualities, late maturity, and persistent broodiness are characteristics of the meat breeds. Brahmas, Cochins, and Langshans, all classified as Asiatic, are the principal meat breeds.

DUAL-PURPOSE BREEDS

These are medium-size, good table birds, fair layers, less active than the egg breeds, but more so than the meat breeds, and are good egg-sitters and mothers. Plymouth Rocks, Wyandottes, and

Rhode Island Reds (all of which are American) belong to this class.

FANCY BREEDS

Bantams of various varieties, Poland, and Silkies come under this heading, and are raised chiefly for some peculiarity of form or feather without regard to any useful qualities.

The above classification of chicken breeds is a loose one (and there are crossovers), but does give a general idea of what to look for when first thinking of raising chickens. When choosing which chicken breed is best for your coop, there is more to consider than simply which chicken will lay the most eggs. If you live in a very cold climate, for example, a bantam breed may not be the easiest to raise, and those with large combs can be prone to frostbite, so more care when insulating and ventilating the chicken coop will be required.

HOW TO CHOOSE THE PERFECT BREED

WHAT IS AN IDEAL CHICKEN?

Your "perfect breed" depends in part on the reason you want to raise chickens in the first place. If you just like the thought of pet chickens pecking around your backyard, you might choose them on the basis of what they look like, rather than their laying ability. If you are thinking about lovely free-range eggs, you will want to choose a breed that lays well. Finally, if you are planning to raise chickens "to show," other particular breeds might be more suitable. Whatever the case, the "best" breed for the purpose will provide you with your ideal chickens.

Chickens come in all shapes, sizes, and colors. There are hundreds of known breeds to choose from. Within the same breed they will share the same skin color, number of toes, and plumage style.

There are two main categories that chickens can fall into: hybrids and purebreds. Hybrids are birds that have been bred especially to lay large numbers of eggs. Pure breeds, some of which have been around for hundreds of years, only tend to lay eggs for part of the year, generally from March to September. The advantage of purebreds is that they are usually hardier and less prone to disease.

Among the breeds, there are further categories. Certain chicken breeds are layers, bred chiefly for egg production, others are bred primarily for table use, which, as a rule, means they have low egg production, and still others are dual-purpose, bred for both egg production and table use. Then there are fancy breeds with special features such as crests, feathered feet, unusual colors, beards, muffs, or five toes.

Some fancy breeds are strictly ornamental, but others will be found among the layers or dual-purpose birds. There are also bantams: smaller chickens that are ideal for people with less space. These chickens originated from the Bantam area of Java, but after they were introduced into Europe, any miniature version of a breed was called a bantam—both those for which there is a large standard breed and those that have no matching standard.

Hens are very sociable birds. They don't like being on their own, so once you commit to raising chickens, you should think about having at least two or three. Cocks, the male of the species, are only necessary if you intend to breed your chickens. Otherwise, they tend to be noisy creatures, and if you have neighbors close by, they may not be happy with the cock's wake-up call at some ungodly hour in the morning.

Most new chicken owners, entranced with the very idea of keeping chickens, will start by having these chickens as pets, with egg production a bonus.

There are several schools of thought regarding breeds. Two of my friends who raise chickens for pleasure and for the eggs have an entirely different rationale for their chickens. One started out with fully grown waifs and strays of no fixed breed and then progressed to specializing in certain breeds by getting them from other friends, purchasing them from markets, and even hatching them herself in an incubator. The other decided on rare breeds right away because, she said, it was a "nice feeling to think that one is perpetuating something that is endangered". Other people I have spoken to have rescued battery hens that might otherwise have been slaughtered. Yet another friend, well-known as a figurative "mender of broken wings," kept a chicken and turkey in her bathroom, for reasons I never understood, and what a surprise I got as I encountered them one evening when I went to powder my nose!

If your aim is to provide your household with eggs, deciding how many chickens to keep (given that you have enough room) is simple. Allow two hens for each member of the family and, with good care, there should be plenty of eggs all year round. Some chickens will lay an egg a day, but usually there will be only three to six per hen per week. These numbers will depend on the weather, the feed, and the strain of bird.

The following are some of the more popular chickens in the USA.

LARGE BIRDS

Araucana
Araucanas are a rare breed of chicken originally from Chile. The hens are good layers of beautiful blue or green medium-size eggs. The eggs are unique in that the shell color goes all the way through. They are not the most prolific layers, producing only about 150 eggs per year.

Barnevelder
The Barnevelder is a pure breed from the Netherlands. Some strains have been bred for show and others for the darkness of their wonderful brown eggs.

Black Rock (Bovans Nera)
The Black Rock is a very successful hybrid containing Rhode Island Red genes and has lush plumage, which enables it to adjust to most weather conditions. It also produces beautiful brown eggs in excess of 300 per year.

Brahma

The Brahma, a large feather-legged, (what I used
to call floppy-footed) bird, was created
in the USA from Chinese Shanghais and crossed
with Malay-type birds from India. The official
name, Brahmapootra, was shortened to Brahma
in the mid 19th century. It lays less than 100
brown-tinted eggs per year.

Cochin

The Buff Cochin is the most widely bred of the
Cochin class, which originally came from China.
It is a very large dual-purpose chicken, laying
about 140 tinted eggs per year, and is one breed
that doesn't come in bantam size.

Leghorn

The Leghorn originated in Italy and developed
extrensively in the 19th and 20th centuries.
It has been bred up from a bantam, is now
classified as a "light" breed, and is a prolific
producer of about 240 white eggs per year.

Maran

Originally from the French port of the same name,
Marans are attractive birds that lay dark brown eggs.
The eggshell is very smooth, with small pores. The egg
white is quite dense, which gives it a good shape in
the pan if you are frying it. Marans produce up to 200
eggs per year. They were previously more common in
Europe but have become popular in the USA.

Orpington

The Orpington originated from the town of Orpington, Kent, in the UK. It is a large, beautiful bird. The Orpington has very thick plumage, which contributes to its size. It is nice and friendly and lays light brown eggs.

Rhode Island Red

The Rhode Island Red, developed from Asiatic black-red fowls of Shanghai, Malay, and Java types, found its way to Massachusetts where it was first exhibited in 1880. Now popular all over the world, it is a dual-purpose, medium-heavy fowl. It is a good layer, producing more than 250 eggs per year on average. It is a hardy bird; very friendly, easy to keep and ideal for first-time or young chicken enthusiasts. Its eggs are a rich brown color.

Silkie

The Silkie, which originated in China about 3000 years ago, comes in white, black, gold, and partridge varieties, and may be everyone's favorite breed. They are perfect pets—friendly, quiet, tame, and just beautiful to look at. They are smallish and only lay 80—100 small eggs per year, but they are excellent brooders for other hens' eggs. While they are quite small, they are classified as "light large" and not "bantam."

Sussex

This bird originated in Sussex, England, and has a long British history, especially the speckled variety. It was one of the main table birds for the London meat market. It is now available in colored, white, or light and is quite a good layer, producing about 260 eggs annually.

Welsummer

The Welsummer is a typical backyard bird, laying 200 large brown, sometimes mottled eggs per year. It was named after the village of Welsum in Holland, but is quite a popular breed in Europe, as it is a good free-range chicken. This breed is rare in the USA, but very common in Oklahoma.

Wyandotte

The Wyandotte is a very pretty bird in a variety of colors. The first strain of the Wyandotte family was the Silver Laced, originally from the USA, where it was standardized in the late 19th century. It was introduced into England about this time, and English breeders crossed it with Cochin and Hamburg males to produce a gold-laced variety. Other colorations were produced from various blendings of breeds. The Wyandotte is a good layer, producing over 200 light to rich brown eggs.

Bantams

As mentioned earlier, many breeds also come in bantam sizes. The Pekin is only available as a bantam.

Pekin

A true bantam, the White Pekin (originally from China, hence the name), is a spectacular-looking hen with snowy white feathers. They make marvellous pets for children as they are easily tamed as well as being good show birds. There are all sorts of colors of Pekin—lavender, black, buff, and speckled, to name some. Pekins produce only a fair amount of eggs per year, 150 on average.

All of the above breeds are readily available from reputable breeders.

CHICKENS IN SMALL SPACES

If you would like to raise poultry, but have a limited amount of space, you can still be a proud keeper of happy chickens. One option is to simply choose smaller birds, such as bantams.

Bantams are miniature chickens available in many of the same breeds as the larger ones as well as some more exotic varieties.

Bantams can be the ideal chicken for the novice poultry keeper and because they are docile and friendly, they are superb pets for families with children.

Bantams eat less than larger chickens and produce smaller, but perfectly lovely eggs. Bantams are often sold as trios—two females, plus a male, but it is not necessary to have a cockerel unless you want to breed them. A good breeder should be able to give you what you want, so look around for one. A trio of bantam chickens will happily thrive in a small portable coop, or even a rabbit hutch that can be moved around the backyard every week to provide fresh ground for them.

Silkies are another good choice for small areas and families with small children, because they have a very sweet nature. The Silkie looks like a large chick and children are entranced with this bird, which is easy to pick up and hold because it has small wings. When larger birds flap their wings, sometimes little children are a bit frightened. Children should always be supervised around birds to ensure that they do not injure them.

Silkies have fine hairlike feathers and lay small, perfect eggs. They will do well on a small backyard plot.

PURE AND RARE BREEDS

Pure-bred chickens are not kept by commercial farmers because their egg production (or meat) is not as good as that of hybrid versions.

If you are thinking about pure breeds for showing purposes, the criteria is different, but if you want to prevent certain strains of chicken from becoming extinct, it is still advisable to source the best and never knowingly settle for substandard birds. Many once common pure breeds are rare, so conserving these breeds is becoming more and more important.

Vulnerable breeds listed by the American Livestock Breeds Conservancy in January 2010 are Ancona, Cubalaya, Dorking, Java, Lakenvelder, Langshan and Sussex.

Among the endangered breeds are Andalusian, Buckeye, Chantecler, Delaware, Holland, Malay, Redcap, Russian Orloff, and Sumatra.

Pure breeds can be separated into two groups—the "light" or Mediterranean group and the "heavy" or Asiatic group.

The characteristics of the Mediterranean group are lighter body weight, which means they will need less food, have flighty, nervous dispositions, are not broody and, as a rule, lay eggs with white shells.

Asiatic traits are heavier body weight, eggs with brown or tinted shells and broodiness.

Some pure and rare breeds you may encounter are:

Andalusian

This is one of the oldest of the Mediterranean breeds and owes its name to the province of Andalusia in Spain. Two sizes exist—large and bantam, both classified as light breeds. The blue-laced coloration does not breed true, but produces 25 percent blue, 25 percent splashed (white with gray splashes) and 50 percent black. Only the blue variety is exhibited, but if black and splashed are mated together, more blues are produced, although the lacing can become a little indistinct. As egg producers, Andalusians compare well with other breeds, as the hens lay a good number of white eggs. They are hardy, robust, energetic foragers, yet very attractive fowl, being active and elegant. The young birds grow and mature quickly.

Campine

The continental Campine is a beautiful golden/silver bird with unique penciling. It has large and bantam varieties. It may have an ancient ancestry, but was refined and further developed in Belgium in the 19th century. It lays white, medium-size eggs. It is an economical eater and is adaptable to confinement, but prefers to be free-range. It is lively and inquisitive.

Houdan

Originally from France, the Houdan was introduced into England in 1850. It takes its name from the town of Houdan and has been developed for its table qualities. It is a heavy breed and is one of the few breeds with a fifth toe. Other distinctive features are the crest, leaf-shaped comb, muffling and beard, and there is only one standard color—black, mottled with white.

Legbar

The Legbar was the first autosexing breed developed in Great Britain from the Mediterranean Brown Leghorn. The most popular cream legbar is crested and lays beautiful bluish-green eggs.

All the new autosexing breeds have names that end in "bar" and begin with the first syllable of the breed from which they were developed. For example, Dorbar, Rhodebar, and Welbar were bred from Dorking, Rhode Island and Welsummer. Autosexing means that on hatch the girls and boys (hens and cockerels) can be told apart (sexed) by the color of their down. Generally the boys hatch yellow with muted patches of brown and the girls hatch brown with clear stripes on their backs.
The girls also have much clearer eye lashes. These birds are all pure breeds and will hatch true—they are not hybrids!

Norfolk Grey

The Norfolk Grey was created by Mr Myhill of Norwich, England, under the original name of "Black Maria," and was first shown at the 1920 Dairy Show. It is a rare breed and almost disappeared in the early 1970s. Both large fowl and bantam versions are standardized in one plumage color and are classified as a heavy breed. Active and foraging, the Norfolk Grey produces tinted eggs and is capable of attaining a good size for meat production if allowed to mature slowly. The breed appears fairly regularly at shows, particularly in its county of origin, and is capable of winning prizes when on top form.

DUAL-PURPOSE BREEDS

If you are not a vegetarian and are not squeamish about eating your livestock (think of it this way, rather than eating your pets), you could raise dual-purpose chickens for the best of both worlds—eggs and meat.

Chicken is a very good source of protein and can make a significant contribution to the nutritional needs of people of all ages. It has many essential B-group vitamins and is easily digestible. If you raise your own, it is utterly delicious as well. Once you taste a truly free-range chicken, you will never be able to eat the other sort ever again.

Dual-purpose chickens don't lay as prolifically as laying hens and don't grow as fast as the purpose-built meat birds, but they lay better than meat birds and grow faster than laying hens.

Dual-purpose chickens share these characteristics:
• They are reasonably large-bodied
• They are usually quite self-reliant and hardy
• Most of them lay brown eggs
• They are good brooders

Dual-purpose breeds are our traditional backyard chickens. Their advantage is that if you have an excess of young males or a hen that is no longer laying, they are large-breasted and have a good quantity of meat on their bones. Pure-bred dual-purpose chickens in the USA include New Hampshire, Rhode Island Red,

Barred Plymouth Rocks, White Plymouth Rocks, and Wyandotte. There have been several hybrid dual-purpose birds developed to satisfy the market for backyard chickens.

The surge in the popularity of keeping chickens, which is almost reaching the scale of the poultry mania the mid-19th century, has been prompted by our demand for healthier food. The availability of purpose-bred varieties which thrive under semi-liberty conditions makes it possible today to produce good eggs in your backyard with very little difficulty. As the *New York Times* puts it: "Scratch a suburb, find a chicken."

No one can actually tell the difference between a white egg and a brown one once you have thrown away the shell, but the latter are seen as more attractive, more "natural." These brown-shelled eggs are traditionally produced by heavy, meaty breeds with red ear lobes. Hybridizers had a tremendous struggle to create a really efficient brown-shell breed.

The good news for contemporary poultry keepers is that geneticists have turned their hand to improving free-range hens, making their hybrids hardier and better feathered, calmer in temperament, and slightly more adaptable in their dietary needs.

There are a number of prolific layers of brown eggs, such as Rhode Island Reds, Black Australorps, Bard Rocks, Golden Comets, and Red Sex Links.

If you keep a dual-purpose, meat-and-eggs breed, a realistic total over the laying cycle of 12—14 months is likely to be 2000—2500 for 10 hens. Each hen will be eating about 5 pounds of food for every dozen brown eggs on the table.

The standard backyard bird used to be listed in advertisements as 'RIR X LS POL', which translates as 'Rhode Island Red cross Light Sussex point-of-lay' pullets. These are not only dual-purpose birds producing eggs and meat, but also easily sexed at day-old: the females of this cross are dark like the male Rhode Island while the (usually) unwanted cockerel chicks have the pale down of the female parent.

If, like some of my friends, you are happy to eat your chickens, secure in the knowledge that they have lived happy lives and been fed only good feed, but are unable to do the actual killing, you should be able to find a kindly farmer who is experienced and will do it for you, especially if you share some of your birds.

EXHIBITION POULTRY

People who keep and breed poultry for exhibition are known as "fanciers," and the hobby is known as "The Fancy." Exhibition chickens can also be kept to produce eggs for home consumption, while some are kept only as pets.

Poultry fanciers will accept less eggs from their pure breeds in return for the pleasure these beautiful birds bring them and for the joy of exhibiting them around the country.

There are several reasons for wanting to show your chickens:
- To get feedback from judges and other exhibitors
- To win awards
- To sell birds
- To arouse interest in a breed or variety nearing extinction
- For the social interaction

How well your bird will place depends on its condition, its disposition, how closely it conforms to the description for its type, how it compares with the other birds on show, and how good the judge is.

Those who win consistently are exhibitors who are also breeders with an in-depth knowledge of the genetics of their chosen breed and variety. People who have gone beyond the idea of chickens clucking away in their backyard and want to take a serious interest in breeding and exhibitions will find *Exhibition Poultry Keeping* by David Scrivener of much interest.

Many exhibition birds are very valuable and are traded strongly. Their owners would also make a strong case for the importance of their genetic material being preserved. All of the pure breeds and varieties or sizes within a breed are rare.

An organization known as The American Livestock Breeds Conservancy, founded in 1977, describes itself as "the only organization in the USA working to conserve rare breeds and genetic diversity in livestock." It classifies livestock according to several categories: "critical," "threatened," "watch," "recovering," and "study."
The chicken breeds classified under the "critical" category are: Andalusian, Aseel, Buckeye, Buttercup, Campine, Catalana, Chantecler, Crevecoeur, Delaware, Faverolle, Holland, Houdan, La Fleche, Malay, Nankin, Redcap, Russian Orloff, Spanish, Sumatra. Breeds classified under the "threatened" category are: Ancona, Cubalaya, Dorking, Java, Lakenvelder, Langshan and Sussex. Those on the watch list are: Brahma, Cochin, Cornish-non-industrial, Dominique, Hamburg, Jersey Giant, Minorca, New Hampshire, Polish, Rhode Island White, and Sebright.

Exhibition poultry shows are very popular in the Midwestern USA. The American Poultry Association

(APA) publishes an illustrated guide to poultry entitled *The American Standard of Perfection*, which provides a complete description of all the recognized breeds and varieties of domestic poultry in the USA. Chickens presented in exhibitions are judged according to this guide's descriptions of ideal breed type,weight, color, and other characteristics of that particular breed.

In most cases, bantams outnumber large fowl at poultry shows. Among the most popular exhibition bantam breeds are: Cochins, Old English Game, Plymouth Rocks, and Wyandottes. Exhibition large fowl breeds are divided into classes, usually based on their origin. Among some of the most popular large fowl breeds are: Black Australorps, Leghorns, Plymouth Rocks, and Rhode Island Reds.

American Poultry Association exhibitions, known as "meets," are sponsored by different states and Canadian provinces each year. The American Bantam Association (ABA) also sponsors many exhibitions, and there are numerous state and regional events throughout the year.

With regard to the sale of birds, a significant number of young birds are sold, but the majority are sold as adults. Adult birds are usually sold as trios (one male, plus two females) or pairs. Poultry auctions and private sales also take place. Specialist breeders will usually only sell directly from home unless they are attending a combined show and sale.

People tend to treat their special poultry pets in much the same way as a much loved cat or dog.

CHOOSING THE BEST AGE TO PURCHASE

Now that you have decided what sort of chicken and how many you would like to raise, unless a kind friend or neighbor has offered you a few hens from their flock or you are planning to give rescue chickens a new lease of life, you should choose the age you would like to start with—whether to purchase newly hatched chicks, "point of lay," or more mature birds. There are advantages to all these options.

NEWLY HATCHED CHICKS

This may not be the best plan for beginners since baby chicks require an awful lot of attention, but these fluffy babies are very cute, and it's a way of getting acquainted with your chickens as they grow. Baby chicks are often picked up a lot so there is a real chance of bonding with your birds. You will also get more baby chicks for your money than older ones and they are unlikely to carry any diseases. However, if you are in this for exhibition purposes, you won't know if you have a winner until much later.

If you do choose the fluffy newborn route, you can either buy unsexed or sexed chicks. Unsexed chicks have not been sorted with regard to gender. Day-old chicks are virtually impossible to sort—they are sold "as they were hatched"—in other words, theoretically, 50 percent cockerels (male) and 50 percent pullets (female). But somehow it doesn't work out that way in actuality, and you are likely to end up with more males. If you want lots of eggs, this isn't the way to get them. Unless you are raising cockerels for meat, most of your chickens should be hens. Too many cockerels will fight with each other and wear your hens out.

Sexed chicks are sorted so you can buy as many of each gender as you want. Usually pullets cost the most, unsexed next and cockerels least. The reason cockerels cost so little is that most breeders have too many of them.

Many types of hybrid chicks can be sexed either by their down color or the length of their wing feathers. This is known as autosexing, but it is not an exact science and is best left to the expert breeders. Pullet chicks will have primarily reddish-brown down while the males are white or cream-colored. Feather sexing is more common among table chickens. Day-old male chicks will have same-size feathers covering the base of the wing feathers, while pullet chicks will have longer covert wing feathers.

The downside of purchasing newly hatched chicks is that you will have to feed and house them for about 20 weeks before they produce even one egg. The upside is that this could be a real labor of love and, in addition, provide sweet little pets for you and your children as the chicks grow.

You should only buy chicks from a reputable local breeder. This way the chicks won't have to travel too far and won't run the risk of getting chilled. You will also have to have a brooder to keep the chicks warm once you get them home. The brooder must be small enough to keep the chicks close to each other to provide warmth.

POINT-OF-LAY PULLETS

While raising chicks can be fun, six months is a long time to wait before the first egg is produced.

Buying point-of-lay pullets is the quickest and most popular way of starting egg production. Pullets are female chickens that have yet to lay eggs. They can be purchased at the age of 18–20 weeks and you can expect the first eggs two to six weeks later, rather than the six months it would take until your day-old chicks are ready to lay.

Point-of-lay pullets should be purchased from breeders specializing in this type of chicken. Local farmers are usually good sources of recommendation. When you buy your birds, you should ask the breeder for details about how the chickens have been reared and for recommendations on future management. It is particularly important to determine that the chickens have been vaccinated and find out what the feeding and lighting regime has been.

YEARLINGS

Commercial poultry producers only keep their hens for one year or so after they start laying, so another possibility is to buy 18-month-old birds, called yearlings. Because their main job has been to lay eggs, they might have started to molt and look a bit worn out—you would also after laying 250–300 eggs! After hens molt they will lay fewer, but much larger eggs. So they still have plenty of life (and egg-laying capacity) in them and would appreciate a less stressful environment, although I wouldn't recommend this option to someone who is a novice.

Rescuing battery chickens takes more thought than just putting them in a big enough space and providing proper food, water and shelter for them. Imagine them, if you will, as children who are products of an austere orphanage. They will need lots of love, supervision and patience from you in order to become happy again. They may find the open spaces frightening and choose to stay inside because life in a cage is all they have known. More information about how to deal with rescue chickens can be found later in the book. I have just included a few words here to give you another option.

I have mentioned earlier that the best method of buying chicks is to purchase them from a reputable local breeder. But this goes for all chickens, really —if you are buying chickens of any age, the best place to purchase them is a reputable breeder.

BRINGING YOUR CHICKENS HOME TO ROOST

Looking after chickens is generally a matter of common sense, but there are a few basic rules. You need to use good quality feed, keep your chicken house and run clean, and ensure that your birds are free from predators.

If you do these things, you can look forward to many years of trouble-free hen keeping, learning to enjoy their different characters, watching their antics, and eventually, enjoying the fruits of their labors—delicious free-range eggs.

Before you bring your chickens home, you have to ensure that everything is ready for them (and that they have a place to roost). Whatever age you choose, you can't just leave them to root around in the backyard, pulling up worms.

The natural predator of the chicken is the fox, so you have to make as certain as you can that the fox won't have your chickens for dinner or kill them just for the fun of it. Hopefully you will know where to obtain feeders and watering supplies; if not, feed shops should be able to advise you.

But first, let's talk about a home for the chickens. Chicken houses are as different as the varieties of chickens you can raise. Basically, the best type depends on the amount of space you have, the number of chickens you want to keep, and the amount of money you want to spend. There are many ready-made chicken houses for sale, some of which are very attractive and some that are very pricey.

If you need to set your sights lower, or are good at DIY, there are many other possibilities. You could adapt an unused shed, dog kennel or rabbit hutch, or go over the top and build a miniature mansion for your hens. In general, if you are building a coop, one that measures 6 feet x 3 feet will be adequate space for up to eight chickens.

The fundamentals of good poultry housing will include the following elements:

- It should be easy to clean
- It should have good drainage
- It should protect your flock from the elements (cold, wind, sun, rain, snow)
- It should keep out rodents, wild birds, and predators, such as foxes
- It should provide enough space for the size of your flock
- It should be well-ventilated, but free from drafts
- It should be possible to maintain a uniform temperature
- It should have a place where your hens can roost
- It should have nesting areas to encourage your birds to lay eggs indoors
- It should have sufficient light
- It should have proper feeding and watering facilities

THE HOUSE

Open housing is easier to clean than a complicated coop. It should be tall enough for you to stand in, which will encourage you to clean it as often as necessary.

The house should have a chicken-size flap (called a pop hole) opening downward to form a ramp, so that your birds can come out and go in when they like, but to keep predators out, it should have a secure latch that you can fasten in the evening after your birds have gone inside to roost. Pop holes should be situated so that opening and closing them every day is not too onerous. They must be secure and foxproof and large enough for the chickens to come through, without being so large that small children can crawl in and out of them—1 foot square is about the right size—and two or three are usually enough. It is safer from predators if the pop hole "doors" open and close vertically rather than slide from side to side.

The house should also have a people-size door that fastens securely so you can enter and exit.

ROOSTING

Wild birds roost in trees. Domestic chickens are usually too heavy to do so and trees aren't always available in a backyard. Chickens need somewhere inside their house to sleep, so you must provide perches for them; about 1 foot per chicken. Some chickens prefer to snuggle up close to each other; some like their own space. The perch needs to be high enough to be free from draughts—about 1—2 feet off the ground. This makes it easy for them to jump on and off without injury.

Perches can be made from wood and the ideal size for a chicken's feet is 1—2 inches square. The wood should be sanded smooth so there aren't any splinters, or if you have an old wooden ladder lying around, you could use that. If you are using new wood, round off the corners so the chicken can wrap its toes around it. Chickens will not fall off their perch even if they are sleeping because of some peculiarity in their legs. When the chicken grips the perch, her muscles and ligaments lock onto it and the grip doesn't loosen even when she sleeps.

The perch needs to be removable so that it can be cleaned every week, otherwise droppings would accumulate, and this is bad for the health of your hens' feet.

BEDDING

Bedding, scattered over the floor of the hen house, offers numerous advantages. It absorbs moisture and droppings, it cushions the birds' feet and it controls temperature by providing insulation. Good bedding is inexpensive, durable, lightweight, absorbent, quick-drying, easy to handle, doesn't mat up, isn't musty or moldy, has not been treated with toxic chemicals, and makes good compost and fertilizer.

Most people agree that the best material for bedding is wood shavings. You can start young birds on 4-inch-deep bedding and work up to double that by the time the bird is mature. Chickens like to scratch around in the shavings and take dust baths. The only time wood shavings are inadvisable is with very young chicks who may eat the shavings thinking that they are food.

DUST BATHS

Out of doors, the best sort of dust bath you can provide for your birds is a large, flat litter box filled with sand and put in a sunny place in your backyard. Watching chickens dust bathe is very entertaining and they seem to get a tremendous amount of pleasure out of both the sunshine and their bath.

Do not leave the dust bath outdoors when it is raining because chickens like dust, not mud.

NESTS

It is very important for your chickens to have a place to lay their eggs. They like to do this in dark, out-of-the way places. Nest boxes encourage them to lay their eggs where you can find them and where the eggs will remain clean and unbroken. You can purchase purpose-built nest boxes, but why bother if you have spare cupboard drawers or wooden boxes? Make the nest cosy and comfortable by lining it with bedding. The nests should be neither so small that the chicken is cramped, nor so large that she doesn't feel safe and secure. A good-size nest, especially if you are making it yourself, is about 1 foot square. This is an ideal size for one chicken.

Initially, place the nests (provide at least one for each two to three hens in your flock) on the ground until your pullets get accustomed to them, then attach the nests 18–20 inches off the ground. This will discourage your chickens from scratching in them and possibly soiling them or breaking the eggs. Broken eggs should certainly be avoided, because once a hen gets a taste for eggs (and they do), they will break the shells deliberately. Raising the nest off the ground will also deter rats who love to eat eggs. If you make the nests cosy and warm and inviting for your hens, they will be happy to lay in them and you won't have to search around for your breakfast every day.

Outside the house, the more space your chickens have, the happier they will be. However, if it is very cold, snowing or raining, they may prefer (as we do) to remain indoors.

If you have a large enough backyard, it is a good idea to move the house or coop from time to time to a fresh patch of grass. If your backyard is too small to accommodate this, you will have to consider what sort of flooring your hen house should have. Wood shavings or chips, or straw are all practical for use as floors, but you will need to change them frequently, so it is probably much more economical to buy these materials from a farm or feed merchant rather than from a pet shop.

It is a good idea to keep your birds confined to their house and run for the first week to let them acclimatize themselves to their new home. Don't be tempted to let them roam too soon. Some young chickens will attempt to go back to their original home like homing pigeons and this is usually unsuccessful because they don't know how to fly very well. So keep the door shut and latched until they become accustomed to being there.

The first time you let your chickens out of their house into the big world (the backyard), do so in the late afternoon. Using this strategy, they won't stray far, preferring to keep their house within sight. It will also be easier to coax them back inside when it is time to put them to bed. In the days when chickens were wild, they roosted in the trees for safety and it's an instinct that tame birds haven't lost. Usually, chickens will take themselves to bed at dusk, so we should just make sure they are home before closing and latching the door. A word of caution, though. If you have more than a few chickens, you should count them and ensure that they are all indoors.

PUBLIC ENEMY NUMBER ONE

Your chickens' worst enemy is the fox, and it is very difficult to keep this predator at bay because there is no telling what time of day he is going to decide to visit. Badgers can also be a problem, but if they get into the henhouse by mistake, they may be more frightened than the chickens.

The best deterrent is a fence. Chicken wire netting 78 inches high is a satisfactory material to use for a fence. Since foxes are good climbers, an additional 2 feet can be added to the height, arranged so that it slopes upward and outward. This is high enough so that the chickens will not fly over it. It should be dug in 1 foot into the ground, since the fox will attempt to dig underneath it. You can also use electric poultry netting, which many people swear by.

Other fox deterrents are dogs, which they don't like much, human hair, and male urine, but a fence is definitely best. However, there is no 100 percent preventative, and as careful as you might be, you will probably lose a few chickens because foxes are extremely clever.

CHICKENS V. YARDS

If you are an avid gardener and allow your chickens to range freely, your backyard will surely suffer. The birds will deposit NPK and destroy weeds and bugs, but they will also eat new sprouts, scratch up seedlings, and peck holes in berries and tomatoes. They just love ripe tomatoes!

There are ways to keep chickens successfully without your backyard going to wrack and ruin.

Any plant that is well rooted and mature can tolerate chickens scratching around its base as long as the chickens have a large enough area to range in and don't linger in one area too long.

Let chickens into your backyard late in the day, giving them an hour or so before dusk (when they will put themselves to bed) to eat bugs and nip leaves, but not enough time to do serious damage.

Keep a breed with heavy leg feathering since they tend to scratch around less than other breeds and won't do as much damage to seedlings.

On the plus side, chickens can do some of your gardening chores for you. They provide natural pest control, eating snails, slugs, earwigs, and other insects. But do be careful and only use non-toxic, organic pesticides on any plants they might eat.

HANDLING, FEEDING AND WATERING

"It's only chicken feed," describes a very unimportant amount of money, but after the initial outlay of making sure your chickens have adequate and safe housing, the cost of feeding your chickens will be the most expensive item you will have to budget for.

Commercial feeders are available in many sizes and are usually either plastic or galvanized steel. The plastic ones are quite inexpensive and easy to clean, but will need protection from the elements so they need to be sited carefully. Galvanised steel feeders, which usually have a cover are much sturdier and will last a lifetime. They are also less likely to be knocked over. Either way, they should be raised off the ground to keep the food clean. Both types are designed so that your chickens can feed without contaminating the food. A tube feeder generally provides for all the needs of your chickens. They keep a small amount of food in the base and as the chickens feed, the tube allows more food to drop into the base.

Choose a feeder that will accommodate the number of chickens you have or anticipate having. Don't be tempted to buy a larger one so that you can put more food in it, because chickens do not tend to overeat and the food will get stale and moldy in a few days, and you'll have to discard it. Feeders and waterers do not need to be placed in the hen house unless the hens have to be confined for some reason because the birds will probably knock them over. They do not feed or drink once they have gone to bed.

If you must put the feeder in the house, it should be hung from the ceiling of your hen house or coop at a height that vermin can't reach, but the chickens can without scattering the food; ideally, chicken neck height. If hanging the feeder isn't practical, the feeder can be situated on a concrete slab. If the feeder has a top, it can be completely filled, but if it is open, only fill it halfway, otherwise the chicken may jump up and feed from the top which will contaminate and waste the food.

You will want to ensure that your potential layers get proper nutrition. Complete feeds for layers are available from feed merchants, but if your chickens are pets as well as for producing eggs, only putting commercial feed in a feeder or trough will not give you any interaction with your chickens at feeding time, which could be an ideal occasion for bonding with your new pets. So you may want to introduce a few "treats" as discussed below.

Once your birds are happy in their new home, you can start taming them so that you will be able to pick them up whenever you want or need to.

Learning to handle your chickens is absolutely crucial on two levels: it aids in the bonding process and enables you to prevent health problems developing. Picking up your birds takes practice, just as picking up newborn babies does. I don't think that either is instinctive. But it is easy to learn how to do it.

Wait until your chicken is feeding. You should be able to gently stroke her back feathers. Once she is comfortable with you doing this, you should be able to pick her up. Another good time to handle chickens is when they are roosting. Chickens become very quiet after dusk and will not struggle if you touch them. Confidence is very important. Bring both hands down around the centre of her body so that you are holding her wings in. Do this gently so that you are not putting too much pressure on her ribs. Next, lift her up and hold her against your body so that one wing is pressed against your chest. Slide one hand underneath her body so that she is supported and place the other hand on top, holding in the other wing. Make sure she can't flap her wings because this will cause her distress. You can either then place her under your arm or keep her close to your chest. Your chicken will feel quite safe and you should be able to stroke her in this position, making her feel even more relaxed.

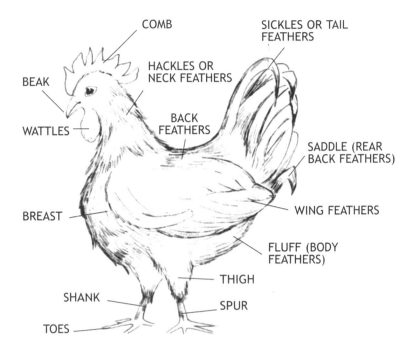

COMB

SICKLES OR TAIL
FEATHERS

HACKLES OR
NECK FEATHERS

BEAK

BACK
FEATHERS

WATTLES

SADDLE (REAR
BACK FEATHERS)

WING FEATHERS

BREAST

FLUFF (BODY
FEATHERS)

THIGH

SHANK

SPUR

TOES

Before discussing what sort of food is best for your chickens, it might help to understand a bit about the chicken's digestive system. I first heard the expression "as rare as hens' teeth" when I was young. I didn't know what this meant, believing that a chicken grew teeth and they fell out through some normal order of things. But birds don't have any teeth! Therefore they cannot grind up their food with their molars like mammals can. They do have sharp projections on their tongues which help them draw food to the back of their throats, enabling it to go on its journey toward their stomach.

Chickens eat food that is mixed with their saliva.

From there it travels down the gullet into a temporary storage receptacle called a crop, which is a large swelling on the side of its chest. This is normal and no cause for alarm. The food, now softened, enters the stomach. Here it is mixed with digestive juices and goes into the gizzard, a thick muscle beyond the stomach that all birds have. This is their primary aid to digestion. Grit and small stones that the chicken has ingested help to grind the food up, ready for further digestion. The food then travels through the rest of the intestines, getting more digestive juices from the liver and pancreas. Nutrients from the food are absorbed by the intestinal walls and the undigested portions of food pass to the cloaca where they mix with urinary waste products and are expelled through the chicken's vent (outside opening of the cloaca).

Like all birds, chickens do not urinate—they have no bladder. The kidney removes waste in the form of a thick, white substance called urates. This is expelled with the droppings through the chicken's vent.

Managing your chickens' food depends on your experience and the type of feed available. There are several different systems to consider, but you might like to know what a laying chicken's nutritional requirements are.

A good complete commercial feed for free-range layers should include a minimum of 16 percent protein. It will probably also include 3.5 percent

oil, 5 percent fiber and vitamins, minerals, and other natural ingredients needed for your hens to produce a good-size egg with a natural golden yolk. It should also be composed of a combination of wheat, barley, peas, alfalfa, soya, and linseed.
An average hen will eat 4—5 ounces per day. Fresh vegetables or mixed corn can still be fed as a treat, but since the hen's crop can only hold about 3 1/2 ounces, it is best to offer them separately, later in the day, or she will be unable to eat a sufficient amount of feed. If grit isn't already in their feed, they will need some as a supplement to help break up the food in their crops—it acts as "teeth." Another nutritional extra is oyster shell, which will help your hens form good shells on their eggs and provide them with calcium. There is some controversy about whether pellets or mash feed is best, but I think pellets win out because they will produce less waste (mash can be messy and dusty), and chickens seem to like pellets better.

If you are starting out with newly-hatched chicks, you should start feeding baby chick crumbs instead as soon after hatching as possible, continuing until the chicks are 6—8 weeks old and can feed themselves. In the week before offering free-access food (via a feeder), you should mix crumbs with the pellets or mash to ensure a gradual changeover.

If you also give your chickens lots of vegetables, put them into a large bowl or basin or hang them somewhere the hens can reach them. Carrots, parsnips, turnips, and rutabagas are ideal because they are easy to hang up and chickens like them. Vegetables such as corn on the cob are rich in vitamin A and darken the egg yolk, turning it a beautiful golden color. Cooked potatoes are also very much liked by chickens as are oranges, one of my hen-keeping friends says. Chickens, like most birds, are also fond of bread. Tear it into small pieces and they will eat it out of your hands. Never offer chickens raw potatoes. Never throw vegetables on the floor of the hen house; instead, put them into a bowl or basin. Further, never feed chickens table scraps that may contain meat, or anything that has been in contact with meat products, as this has been seen as a likely cause of salmonella and is now against the law.

Fresh, clean water should always be available for your chickens. Choose a container that is large enough to provide water for two to three days, but not much longer, as the water will get contaminated or develop algae. Plastic fountain waterers are popular and usually have built-in legs, but can be difficult to fill and do encourage algae. Galvanized waterers are heavier, will keep the water fresher and, like feeders, will last a lifetime.

Waterers should be raised off the ground to keep the water clean. One hen will drink about 8 fluid ounces of water per day, a bit more in hot weather, so a 4.8-pint container would last six hens about two days.

It is both useful and fun to figure out how much feed your little flock is consuming. Begin keeping records of the amount, type, and price of feed from the first day you have your chickens. Also note the number of chickens and their ages.

A simple chicken-feed form such as this can be kept.

CHICKEN FEED RECORD

SOURCE OF FEED
Date of purchase:_____
Animal feed store, pet store, mill, or company where purchased:

_____

Address: _____
Phone: _____
Website: _____
Email: _____
Contact person:
TYPE AND INGREDIENTS OF FEED
Type of feed (starter, grower, layer, supplement):

_____

Form of feed (mash, pellets, mixed grains):

_____

My name or code for this type of feed:

_____

Ingredients in this feed (or attach a label to this page):

_____

_____

_____

QUANTITY AND PRICE OF FEED

Weight of feed purchased: _____

Volume (gallons) of feed purchased:

_____

Price of this batch of feed: _____

Feed remaining on hand before purchasing this feed

Type: _____ Volume: _____ Weight: _____

Type: _____ Volume: _____ Weight: _____

Type: _____ Volume: _____ Weight: _____

CHICKENS THAT THIS FEED IS FOR

Number of chickens that will consume this feed:

_____

Breed(s): _____

Age of these chickens: _____

Quantity of eggs this flock produced in the last 7 days:_____

Number of hours per day that these chickens go free-range: _____

Where these chickens were purchased (farm, pet store, or company name):_____

Address of this shop, company or person:

Contact person: _____

Date these chickens arrived at my location:

_____

Notes:

_____
_____
_____
_____
_____
_____
_____

MIXING YOUR OWN FEED

All the experts say that the best way to make sure that your chickens get a nutritionally balanced diet is to buy the right commercial blend. How you feed your chickens depends on their type, purpose, age, the time of the year, and the availability of forage.

In most commercial feeds, corn supplies carbohydrates and soya supplies protein. Rations for chicks contain a high amount of protein. As your birds grow, they will gradually need less protein and more carbohydrates.

Mixing rations is a very complex aspect of poultry management and isn't something you should undertake if you are just starting out. However, if you are thinking about formulating your own rations you must consider the following aspects: the availability of appropriate feedstuff; analysis of the composition of feedstuff; a knowledge of the nutritional requirements of chickens; and the ability to mix feed in a quantity that your flock will be able to consume within a month.

You can purchase all of the ingredients separately or even grow some of them yourself. You will need to include carbohydrates for energy (corn and/or other grains), protein for growth (a combination of ingredients that will together supply all the essential amino acids needed for a complete protein), and vitamin and mineral supplements.

Analyzing the composition is probably the most daunting aspect since it will depend on the source, the time of year, and where it was grown. Trying to get small amounts of feedstuff analyzed is probably not at all economically feasible. There are books available that will help you, but they also require prior knowledge on your part.

Then you'll have to consider how you are going to mix your feed. Can you find a local mill to do it for you, or will you need to buy a mixer?

Having given you all the disadvantages first, the benefits of formulating your own feed give you control over what your chickens eat, and may even save you some money (although the amount might be so small that it wouldn't figure in the consideration).

If you have gotten this far and you are still interested, this is what you would need to make up 100 pounds of feed:

STARTER FEED
Coarsely ground grain (e.g., corn, milo, barley, oats, wheat, rice)......................... 46 pounds
Wheat bran, rice bran, mill feed, etc. .... 10 pounds
Soya bean meal, peanut meal, cottonseed meal, sunflower meal, sesame meal, etc. .... 29$\frac{1}{2}$ pounds
Meat meal, fish meal, soya bean meal .... 5 pounds
Alfalfa meal (not needed for free-range birds) .... 4 pounds

Bone meal, defluorinated dicalcium phosphate..2 pounds
Vitamin supplement (supplying 200,000 I.U.
vitamin A, 80,000 I.C.U. vitamin D3, 100 mg riboflavin)
Yeast, milk powder (not needed if vitamins are
balanced) ....2 pounds
Ground limestone, marble, oyster shell............1 pound
Trace mineral salt or iodized salt supplemented with:
  0.5 ounces manganese sulphate and
  0.5 ounces zinc oxide ....0.5 pounds
TOTAL..100 pounds

GROWER FEED

Coarsely ground grain (e.g., corn, milo,
barley, oats, wheat, rice)..........................50 pounds
Wheat bran, rice bran, mill feed etc. .... 18 pounds
Soya bean meal, peanut meal, cottonseed meal,
sunflower meal, sesame meal, etc. .... $16^1/_2$ pounds
Meat meal, fish meal, soya bean meal........... 5 pounds
Alfalfa meal (not needed for free-range
birds) ....4 pounds
Bone meal, defluorinated dicalcium phosphate... 2 pounds
Vitamin supplement (supplying 200,000 I.U.
vitamin A, 80,000 I.C.U. vitamin D3, 100 mg riboflavin)
Yeast, milk powder (not needed if vitamins
are balanced) ....2 pounds
Ground limestone, marble, oyster shell ....2 pounds
Trace mineral salt or iodized salt (supplemented with:
  $^1/_2$ ounce manganese sulphate and
  $^1/_2$ ounce zinc oxide ....$^1/_2$ pound
TOTAL..100 pounds

LAYER FEED

Coarsely ground grain (e.g., corn, milo,
barley, oats, wheat, rice). . . . . 53$\frac{1}{2}$ pounds
Wheat bran, rice bran, mill feed etc. . . . . . . 17 pounds
Soya bean meal, peanut meal, cottonseed
meal, sunflower meal, sesame meal, etc........15 pounds
Meat meal, fish meal, soya bean meal. . . . . . . 3 pounds
Alfalfa meal (not needed for free-range
birds). . . . . 4 pounds
Bone meal, defluorinated dicalcium
phosphate . . . . 2 pounds
Vitamin supplement (supplying 200,000 I.U.
vitamin A, 80,000 I.C.U. vitamin D3, 100 mg riboflavin)
Yeast, milk powder (not needed if vitamins
are balanced) . . . . 2 pounds
Ground limestone, marble, oyster shell . . . . . . 3 pounds
Trace mineral salt or iodized salt, supplemented with:
  $\frac{1}{2}$ ounce manganese sulphate and
  $\frac{1}{2}$ ounce zinc oxide. . . . . $\frac{1}{2}$ pound
TOTAL . . . . 100 pounds

Taken from *Feeding Chickens*, Suburban Rancher, Leaflet
no. 2919, University of California

HOW MUCH TO FEED

The amount a chicken eats varies with the season
and temperature as well as the bird's age, size,
weight, and rate of lay. Since they tend to eat to
meet their energy needs, it also depends on the
feed's energy density. Chickens that are fed the
same rations all year round will eat more in the

cold weather because they need more energy to keep warm.

If a chicken doesn't get enough to eat, it will not thrive or lay well. Chickens sometimes eat too little if they are going through a partial or hard molt, if they are low in the pecking order, or if they don't like their food. Chickens are particularly fussy about texture—they don't like dusty or powdery food.

If you think your chickens are underfeeding, perk up their appetites by feeding more often and giving them variety—chickens particularly like milk, cottage cheese, tomatoes, and salad leaves, all of which can be safely fed in small amounts.

If the chickens are eating too much, this may be a sign of worms, so take a sample of droppings to your vet and worm your chickens if necessary.
If the feed is going too quickly in the winter, it is usually because your chickens are cold. Increasing the carbohydrate intake should help.

Disappearing food might not be the fault of dietary or heat deficiency at all. Other animals and birds might be helping themselves.

FEEDING METHODS

Basically there are two ways to feed your chickens: free choice or restricted. Free choice involves leaving feed out at all times so that your chickens can eat whenever they wish. This is easier to do if you are using a trough, as it means you only have to fill the trough infrequently. The disadvantage is that food is always available, not only for your chickens, but for wild birds, rodents, and whatever wants it.

Restricted feeding allows you more time with your chickens since you are interacting more often, but chickens are very greedy and eat quickly, so some poultry keepers believe that restricted feeding can lead to cannibalism.

SUPPLEMENTS

Chickens that are fed commercial rations usually do not need any supplements, but range-fed or home-mix-fed chickens may need supplements of grit, calcium, phosphorus, or salt.

MOLTING

Shorter days are a signal to birds that it is time to renew their plumage in preparation for migration and the coming winter. Like all birds, chickens lose and replace their feathers at approximately one-year intervals. This process, called molting, takes place over a period of weeks, so that a bird never actually looks completely naked.

Don't panic when your chickens start to molt. This is completely normal, and they won't go bald (although some do look rather plucked). Nature has a way of taking care of things. By the time winter comes, they will have enough feathers to keep warm.

Under typical conditions a chicken molts for about 14–16 weeks during late summer and early autumn. The best layers molt late and fast; the poorer layers start early and molt slowly. The nutrients needed to produce eggs are channeled into plumage production. As a result, most pure breeds stop laying for about two months, while hybrids will slow down but not stop laying entirely.

Feathers are usually renewed in sequence starting from the head and progressing toward the tail.

Molting can also result from disease or the stress of a chill, or going without food or drink. Stress-induced molts are usually partial and are not always accompanied by a loss of production.

During a normal molt, feed requirements, particularly protein, are high, but it is not a good idea to make drastic changes in the feed. A little extra grain may be given when the molting is under way, but otherwise you should continue with the same routine. Handle the birds regularly to ensure that they are in good condition, but some birds don't much like being held when they are molting. Perhaps their spiky new feathers hurt when they are compressed. Once the chickens come through the molt, feed them well and allow access to grit or oyster shell in preparation for another season of egg laying.

CONTROLLING RODENTS

Unfortunately, if you have chickens, you are extremely likely to have rats as well. They thrive on chicken feed and must be controlled or they will become a very serious problem.

One method is to also have cats. But don't expect the cats to survive on what they hunt. They have to be taken care of—fed and watered as you would any other pet, and you should also worm them regularly if they are to be ace rat catchers. To catch the maximum number of rats, your cats must be fighting fit.

Rats reproduce very quickly from the age of four months onward. The gestation period is about three weeks and females can produce upwards of five litters a year with 10 babies per litter. So if you do the math, you can end up with a multiplicity of rats, who will happily eat their way through your chicken feed.

There are only two other ways to get rid of rats that I know of—shooting them or setting poison traps. Shooting is quicker and, in a way, more humane, but I think the person with the shotgun has to be a sharpshooter because rats move very quickly. Poison isn't very nice and can be dangerous or even fatal to dogs and cats, not to mention children, but usually it is the only viable solution.

Rodent poison can be purchased as a ready-to-use anticoagulant solution for you to set the traps, or you can use professional pest-control people who will do it for you. If you do it yourself, you must be extremely careful to follow the instructions to the letter, place it where pets and livestock won't eat it accidentally, and store it locked away in a safe place, out of the reach of children.

Dead rats should be disposed of on a regular basis to prevent other animals from eating them, and also because they attract flies and maggots.

So now that you have prepared against rats eating the chicken feed and foxes eating the chickens, you can look forward to bringing up your flock.

HEALTH, DISEASE, AND VACCINATIONS

Poultry flocks are susceptible to a wide range of diseases caused by various infectious microbes including bacteria, viruses, and mycoplasmas. Vaccination of chicks continues to play an important role in poultry protection, especially for diseases such as Marek's disease, Newcastle disease and infectious bronchitis.

When purchasing chickens, always make sure that you buy them from a reputable breeder. The best source is through personal recommendation, but you can easily find a reputable breeder through the internet. Poultry should be sold with all necessary vaccinations.

However, if you breed your own chickens, you have to decide whether or not to inoculate your birds.

Newly-hatched chickens have a certain amount of natural immunity, and they continue to acquire new immunities as they mature. Ask your veterinarian or poultry specialist to help you decide, based on diseases occurring in your area. Your chickens should be vaccinated only against diseases they have a reasonable risk of getting, including past diseases that have occurred in your flock or new diseases that pose a serious threat in your area. Do not vaccinate against diseases that do not endanger your flock.

Healthy chickens should be your utmost concern and there are various things you can do to ensure a healthy flock.

CLEANLINESS AND HYGIENE

Although it seems redundant to say this, cleanliness is an absolute necessity for healthy chickens. Prevention is much better than cure. Safeguarding your chickens' health is dependent on good sanitation. Slapdash sanitation is the most frequent cause of failure in raising chickens.

Good sanitation includes frequent cleaning of your hen house, feeders, waterers and other equipment, and regularly scheduled clean-up of the backyard. A routine once a week clean should be sufficient.

Remove all movable feeders, waterers, perches, and nests. If you are removing the bedding, you can shovel it out with a garden spade. If you are leaving the bedding in place, shovel it back from the walls. Remove the chicken droppings from perches, walls, and nests with a hoe. Use an old broom or vacuum cleaner to remove dust and cobwebs. Disinfect perches, nests, and the inside of the hen house by adding 1 tablespoon chlorine bleach to 7 pints of boiling water.

Your aim is to keep the bedding dry at all times. You may be able to find a gardener to remove unwanted chicken droppings and bedding.

Recycling the droppings will help you avoid recontamination. Poultry droppings and bedding aren't necessarily a nuisance. They are a valuable commodity that can be composted for gardening.

Keeping chickens in portable housing is one way to avoid a droppings problem. Simply move the accommodation often enough to keep droppings from accumulating. In stationary housing, bedding absorbs much of the droppings moisture, minimizing both odor and humidity.

EASY HEALTH CHECKS

• When fully grown the chicken should sport a nice firm comb. The comb will be bright red when the chicken is in lay.

• The eyes should be beady and bright.

• A healthy chicken will be perky, lean, and active.

• Scales on the legs and feet should be smooth and not lifting.

• The color of the legs is a good indicator of whether the chicken is laying. If they are very yellow then she is probably not laying eggs yet. If they are pale, almost white, then she probably is.

• When you pick your chicken up, her body should be plump and firm, but with no flabbiness.

- You can examine your chicken's eyes and nose to check there are no discharges.

- The vent (the chicken's all-purpose exit point) should be moist and white, with no lumps, crustiness, or bleeding, etc.

POSSIBLE AILMENTS

Chickens can have minor ailments, like colds, but these are easily treated with a bit of good old-fashioned tender loving care.

Most problems occur when too many chickens are being kept together or are being neglected. As long as you have been following the suggested day-to-day and week-to-week routines it is unlikely that you will have any serious problems.

Your chickens may get:
WORMS: Round and tape. These are the most likely types of worms. The symptoms are a drop in egg production with an increase in hunger. Birds can also have diarrhoea as a symptom, although diarrhoea alone does not mean worms are the culprit. Contact a vet who will give you medication to be included in your bird's feed.

LICE: The symptoms are a white build-up around the feather base near the vent. In a bad case there could be a build-up on feathers as well. The whiteness is lice eggs. If you do find lice eggs around the vent when checking the chicken's health, brush them off and rub petroleum jelly around the area. To prevent lice, dust the nest box with louse powder every week or so.

RED MITE: The symptoms are a decrease in egg production. In a bad case the bird could look a bit pale from blood lost to the little suckers.
You will not be able to find any by inspecting your birds during the day as they only crawl onto the birds for a nibble at night. Have a look every month or so for mites in the crevices at the ends of the roosting bars. There are mite-sprays available to prevent this.

NORTHERN MITE: In bad cases, your chickens will have a scabby comb, face, and wattles.
The mites tend to gather around the vent and are gray/black. Again, mite-sprays are available to stop them.

It is very unlikely that your chickens will develop any of the following problems.

INFECTIOUS BRONCHITIS: There will be a decrease in egg production and thin, rough, and wrinkly egg shells. Your birds will sneeze and gasp and have a discharge from the beak.
Your chickens are unlikely to catch this airborne disease as they should have been vaccinated before you got them, but it is still possible. The illness should only last for a couple of weeks. Egg production will improve again, but will probably

never be as good as before, with more occasional dodgy eggs.

ASPERGILLOSIS: This ailment makes your chickens thirsty, wheezy, and lethargic. Avoid this by making sure you clean the droppings tray regularly. Young birds are most vulnerable, but adult birds can also be infected. It is caught by breathing in spores from moldy matter. Unfortunately there is no cure! The best thing for the chicken would be to put it out of its misery, or get a vet to do it for you.

SCALY LEG: Your chickens will have scaly legs. The mites live on the legs under the scales which start to lift up and can give the bird trouble walking. You are probably best off asking a vet about this problem. He will put something on the legs, which effectively suffocates the mites. Dipping their legs in old engine oil is messier, but also does the trick since it drowns the mites!

Where do these illnesses originate?
Chickens can catch infections from wild birds, and to prevent this you can add a natural antibiotic to their water. It is called Citrocidal and is made from grapefruit seeds. Available from health food shops, it is very effective at clearing up any respiratory difficulties such as sneezing, coughing, or rasping breathing.

If the chicken's manure is runny and yellow, this is a sign of such an infection and should be treated with this antibiotic; the symptoms should clear up in a week. You can continue eating the eggs, because Citrocidal is 100% natural.

FEATHER LOSS

When a chicken is about a year old, she will start to lose her feathers, but don't panic, this is normal. She is molting. This is a completely harmless process of plumage rejuvenation. Make sure the birds are well fed during this period, as it takes a lot of energy to grow new feathers. Because of all the energy taken up by molting, your chickens will stop laying until their new feathers have grown. It is also important to remember not to clip your chickens' wings when they are molting.

CLIPPING YOUR CHICKENS' WINGS

If your chickens are totally free-range and you don't have any fences enclosing them, it would be cruel to clip their wings, because if they couldn't fly, they would have no means of getting away from their predators.

On the other hand, if you do have fences, you really should clip their wings so that they don't fly away. This will keep your chickens safer and will help protect the vegetable and flower beds in your backyard from being trampled and eaten.

Clipping, the most common method of controlling the flight of backyard chickens, is entirely painless if it is done properly. Clipping with sharp scissors should only be done to one wing.

You should hold the chicken gently and span out her wings. Cut about 3 inches from the first ten flight feathers. These secondary feathers are not required to keep the hen warm, but will cause her to lack the balance needed for flight. Note that this will only last until new feathers grow during the next molt, which could be a duration of several months up to one year. One potential problem is that clipped feathers may not readily fall out during the molt, which will require your assistance.

Clipping wings is a bit like cutting finger and toenails. You wouldn't want to cut your nails too far down because it would cause pain. Similarly, you don't want to cut too deeply into the wing, which will cause your chicken pain and make her bleed. Before you attempt to clip wings for the first time, it would help if a vet or experienced poultry person showed you how to do it correctly.

WOW! THE FIRST EGG

If you started with newborn chicks, you will have to wait 20 weeks or so until your chickens start laying. If you purchased "point of lay" birds, you may still have to wait a week or two until they get accustomed to their new surroundings.

Each pullet has the potential to produce several thousand eggs through its lifetime, and a healthy hen should lay 200–250 eggs during her first year, depending on breed, and go on laying for 10 years or so. Occasionally you will hear of an aged biddy laying until she is 20, but this is very rare. Just remember that the healthier and happier your hens are, the more eggs they will lay.

As a general rule, hens with white ear lobes lay white eggs and hens with red ear lobes lay brown eggs. An exception to the red ear lobe rule is the Dorking, which lays white-shelled eggs, and the Araucana, which lays blue-green eggs. Also as a general rule, Mediterraneans lay white eggs, Asiatics lay brown eggs and other classes are a mixed bag!

Sometimes you will be lucky enough to find double yolkers, some of which have a ridge around the center of the shell. I am always delighted with these "twins," but my butcher who sells local eggs tells me that some of his customers aren't at all pleased. I can't imagine why, because in my experience it is not only a happy lottery, but a case of getting two for the price of one. Double yolkers may be laid by a pullet whose production cycle is not yet synchronized, or by a particular breed, often as an inherited trait.

Very rarely an egg will contain more than two yolks. The greatest number of yolks found in one egg is nine. Record-breaking eggs tend to be multiple yolkers. To date, the largest hen's egg was laid in Cuba in July 2008. It weighed almost 6$\frac{1}{2}$ ounces. The previous record in the *Guinness Book of Records* was 6 ounces.

The first time you find an egg that one of your chickens has laid, it will be almost as if YOU are the mother and have created this miraculous object! Don't be disappointed if the egg is small. All pullets lay small eggs at first and they don't even lay them every day—maybe two or three a week. But by the time your chickens are a couple of months older, the number and size will be more predictable.
A good laying hen produces about 20 dozen eggs in her first year. When she is 18 months old, she will take a break to molt, then go on to produce larger eggs, but fewer of them; perhaps an average of 16—18 dozen.

Hens sometimes stop laying in the winter, not because the weather is cold, but because the days are shorter. They will provide you with breakfast again in the spring.

COLLECTING YOUR EGGS

It is very difficult to determine when exactly your hens will lay. To begin with you should check to see if there are any eggs first thing in the morning.
It takes 25—26 hours for the egg to be laid, so as the week goes on, whatever time the first egg

appeared, it will be one hour later each day in the week and there will probably be a day or two when your hen takes a break and doesn't lay an egg.

Collect eggs often — preferably two or three times a day. If you work, you should check for eggs before you leave home and on your return so that the eggs don't have a chance to become damaged. It only takes one broken egg for your chickens to discover that they like the taste and they may take to breaking the eggs deliberately to get at the contents. If this does happen, don't panic! Try to find your cannibal hen and trick her by filling an egg or two with mustard after blowing out the normal contents. If you can't find the perpetrator, leave the mustard-filled egg in the nest box. This should alleviate the problem.

Place your eggs in a basket close to each other so that they do not bang together or roll around and crack. Write the date you collected them on the egg with a pencil so that when you store the eggs, you will use the earlier ones first. Pencil the date very gently so as not to break the shell.

If your nests are well-designed and managed, your eggs should be clean when you collect them. A slightly dirty egg can be brushed off. A really dirty egg may be washed in water that is slightly warmer than the egg (cool water may force bacteria through the shell into the egg). Don't get into the habit of routinely washing eggs since water removes the natural bloom that helps protect freshness.

Regarding storage, it is better to place them rounded end up with the pointed end facing down into the egg boxes. I store my eggs at room temperature, but then I never have more than a dozen or so at a time and I usually cook or bake with them within a week or two.

If you are storing larger quantities of eggs, you will either have to store them in a cellar or in the refrigerator — the ideal temperature is said to be 40°F.

Sometimes free-range chickens will lay their eggs outside the hen house in extraordinary places you might not expect to find then. When this happens, the age of the egg may be in question.

There is quite an easy way to determine the freshness of an egg. Fill a bowl with plain water and submerge the egg.

A fresh egg will sink to the bottom of the bowl and lie there horizontally.

An egg that is one week old will sink to the bottom of the bowl, but the rounded side of the egg will be slightly elevated.

A two- to three-week old egg will also sink to the bottom of the bowl, but will be vertical, with the pointed side down. These older eggs will be perfectly all right to bake with, although you might not want to boil, poach, or fry them. Before you use them, however, you should crack them into a dish and smell them. If they smell sulphurous, that tell-tale "rotten egg smell," discard them immediately. If they smell all right, examine them. A fresh egg has a cloudy, firm white (albumin) that holds the yolk securely. A stale egg has a watery, clear albumin that spreads out thinly around the yolk. While this is not always the case, the older the egg, the greater the likelihood that the yolk will break when you crack it into a pan. But this is by no means foolproof as even the freshest egg will, on occasion, have a watery white or a broken yolk.

A really stale egg will float to the top of the bowl and should not be used for any purpose, except perhaps for throwing at politicians!

BREEDING, NURTURING, AND HUMANE KILLING

BREEDING CHICKS

If you have only hens, you will have only eggs. But introduce a cockerel into your flock and the story ending will be very different.

The cock mates several times a day, so if you have too many, you will exhaust your hens. If there are a number of bald patches on your hens' heads and backs, there is too much mating going on. You will either have to increase the number of hens or decrease the number of cockerels. If you are thinking about breeding your own chickens and you have a very small flock of hens, one cockerel will probably be enough. The smallest amount to consider should be a trio—one cockerel to two hens.

Spring is probably the best season to think about breeding since it is the natural mating season, but also because the weather will be most appropriate for raising the youngsters.

It is a good idea if the cockerel is the same breed as the hens. This will, in the long run, provide you with offspring that will give you the best quality and quantity of eggs.

Introduce your new cockerel to your hens at dusk when everyone is likely to be calm. Hopefully there will not be any altercations and you can leave the hen house and let your chickens mate in peace without you egging them on. (Sorry, I just couldn't resist that!) After mating, the eggs laid by the females will usually be fertile for a week.

When you are sure the eggs are fertile (this comes with experience), you can begin to collect the eggs you wish to hatch. Do not choose eggs with very thin shells as they may not be able to bear the weight of the broody hen. Don't pick very large eggs either as they might be double yolkers and twins are not easy to hatch.

BROODY HENS

SIGNS OF BROODINESS

Just because a hen is sitting on a nest, this doesn't necessarily mean broodiness. She may just be resting after having laid an egg or be sitting there for some entirely different purpose.

For some reason, people who keep chickens have been led to believe that broodiness is a "bad" trait. But more and more, they have discovered that natural incubation of baby chicks is one of the true joys of keeping chickens. It is not necessary to hatch a large number of eggs—a broody hen will be just as happy to sit on two or three. Even if you are not looking for replacement chickens, some of your poultry-keeping friends might be delighted to have a couple of new babies. The only downside to hatching your own chicks is the ratio of hens to cockerels, which tends to be about 50:50. If you are hatching a dozen eggs, for instance, you will probably end up with half a dozen cockerels. Aside from the fact that cockerels will run around and chase your hens and wear them out, they are noisy and unless you live in a very remote place, your neighbors will not be well pleased! If you do end up with too many cockerels, there are several avenues you can explore: you can advertise in a local newspaper, put a sign in a nearby pet store or canvass your fellow poultry keepers. You can also take them to a market to sell.

As a general rule, the light Mediterranean breeds, such as Leghorns, tend to be non-broody or unreliable brooders, while the heavy breeds, such as Orpingtons, Sussex and Silkies, are terrific. The best laying breeds, therefore, are probably the least likely to set.

You can't encourage a reluctant hen to set by putting her in an enclosed place with a bunch of eggs. She'll probably break or scatter them in her haste to attempt to get out. But you can find out whether you have a potential broody by putting half a dozen china, wooden or plastic eggs in a nest and seeing if one of your hens shows an interest. If she does, you can replace the fake eggs with real ones.

You can test a hen for broodiness: gently reach behind her and remove any eggs you find. If she squawks and runs off, she isn't the least bit interested. But if she pecks your hand, puffs out her feathers, or makes growling noises, you might have found your incubator. She will probably start setting within a few days.

Clucking is another good sign. Some hens don't start clucking until the eggs are ready to hatch, but some start as soon as they begin setting.

It is a good idea to separate a setting hen from the rest of the flock because she will leave the nest for a few minutes every day to eat and another hen may come into her nest while she is off eating, causing her to move somewhere else. Or when the chicks hatch, other hens may kill the downy intruders in the same way that they might kill a mouse.

If you have more than one hen brooding at the same time, keep them apart from each other or you will have problems with incubation times (if the hens accidentally switch nests) or eggs becoming chilled (if both hens attach themselves to the first chicks that hatch) or one hen taking over the feeding and watering station, keeping the other hen's chicks away.

Move the broody hen to her new home after dark or she will try to get back to her old nest.

Broody hens often have lice, which have to be dealt with or they will be passed on to the new chicks. To treat this, dust the birds under each wing and in the feathers around the vent with an insect powder.

In the beginning you should take the broody hen off her nest every day at about the same time to allow her to feed and drink and get a little exercise. Twenty minutes or so should be enough before letting her return to the nest. After a few days, you can probably allow her to leave the nest at her own pace, but keep an eye on her. Some potential mothers can be so focused that they will sit for days without food or drink and this is not healthy.

Broody hens eat less than layers. But make sure you put food and water near the nest. Most broody hens will hold their droppings until they are off the nest. To prevent accidents, feed your broodies scratch grain instead of their usual laying ration.

When you handle a broody hen, be very gentle. First raise her wings. She may be holding an egg between her wing and her body and it might break. It might break other eggs as well, so be very careful. Eggs might also break if your hen is trying to cover more eggs than she can comfortably handle. One large hen can usually sit on 12–18 eggs of the size they normally lay, and they don't mind if the eggs come from another hen.

The incubation should take about 21 days. After the 16th day, don't disturb either the hen or the eggs. Make sure she has plenty to eat and drink and be patient. When the chicks start to hatch, be watchful, but don't interfere unless you have to. Once the chicks start to peck their way out of their shells, cover the broody hen up and let her get on with it.

The chicks should all hatch within a day of each other. Then the hen will normally remain on the nest for another day or two before introducing her brood. If hatching is very slow, the hen may leave the nest to follow the first chicks, leaving the remainder of the eggs to chill and die. If this happens, gather up the chicks and look after them until the hatch is complete. Sometimes the hen will get so hysterical when she sees the new chicks that she'll attack them. Again, be ready to rescue the chicks and look after them yourself. For this you will need an appropriate brooder to keep the chicks warm. You should be able to get good advice about this from your poultry breeder.

The broody hen should be confined to her coop for the first week after the eggs have hatched, but the young chicks can be allowed out onto the grass if the weather is suitable. The reason for this is to prevent the mother from straying too far away while the chicks are very young.

If your chickens are truly free-range, you may be in for a surprise because a hen might just collect a clutch of eggs in a secret place, so the first you will know about her motherhood is when she appears with her new brood.

Just a word about feed: you should feed the chicks starter feed and put the hen back on layer pellets, placing it high enough off the ground so that the chicks can't reach it, because the large amount of calcium in layer feed will damage a tiny chick's kidneys. Place the starter feed in a chick feeder with holes small enough so that the hen can't get into it and anchor it in some fashion so that it can't be overturned while the hen is teaching her chicks to scratch. Also ensure that you provide the chicks with a water container the hen can't turn over.

RESCUE CHICKENS

Pity the poor battery hen. Her life is devoted to laying eggs, then when she is no longer productive, she gets slaughtered—at the tender age of 12 months. There are approximately 300 million laying hens in the USA, of which about 95 percent are kept in wire battery cages too small to allow them to spread their wings or engage in other forms of normal instinctive behavior.

We have talked about free-range chickens, but what constitutes an organic chicken?

Organic chickens are reared on an organically registered feed based on cereals and vegetable protein. The birds are given no animal products, routine drugs, growth promoters, antioxidants, or any other additives. Nor are they allowed to be fed on feed which contains genetically modified feedstuffs.

Recently, an increasing number of people around the world have campaigned against the brutally short life of the typical egg-laying hen, and to improve conditions for chickens and other poultry animals. The European Union has agreed to abolish the use of battery cages by 2012, and the Humane Society of the United States (HSUS) and other animal-welfare organizations are lobbying for a similar law in the USA. A number of states and communities in the United States have already passed or are considering similar laws, and some fast-food companies and supermarket chains have agreed to enact new animal-welfare policies.

Non-profit organizations such as United Poultry Concerns, based in Vermont, have documented the suffering of battery hens and seek to raise public awareness of the treatment of poultry in the USA and abroad. There are also numerous animal welfare and rescue organizations across the USA that are involved in the rescue of battery hens and other animals and their relocation in animal shelters.

Hugh Fearnley-Whittingstall, the British TV chef, also wants chickens to be able to lead a decent life before they become dinner. Over the summer of 2007 he worked as an industrial chicken farmer. During that time the farm raised 2332 standard chickens and 1654 free-range birds next door in a matching shed. He says that "for the first week you couldn't tell the difference between free-range and standard, but very quickly the standard chicks put on weight and became noticeably more sluggish." When the chicks were just under three weeks old, they opened the pop holes in the shed for the free-range birds, "who quickly learned to do what comes naturally to chickens — pecking grass, eating bugs and exploring. The birds next door were completely confined," he said, and it made a depressing contrast.

If it takes celebrities to make us wake up to the facts of life, so be it. This is all good news for both our chickens' well-being and our good health.

LOOKING AFTER YOUR HENS WHEN THEY STOP LAYING

Eventually, there will come a time when your hens stop laying eggs. Usually this will be around age four to five. Unless you are in this enterprise for profit, or "egg money," you will probably "retire" your pets who have served you so well over the years.

There are various approaches to retiring hens. My lawyer's wife, who keeps her rare breed chickens as pets and egg layers, maintains her older hens as "senior citizens" and doesn't begrudge feeding them until they die a natural death. My husband's god-daughter keeps a single cockerel and a variety of hens strictly as layers, giving any extra cockerels and retired hens to a neighbor, who enjoys their company.

KILLING YOUR CHICKENS HUMANELY

Chickens intended for eating should be taken off their feed for about 12 hours before slaughtering them to allow the crop and intestinal tract time to empty. A full digestive system increases the chance of contamination of the carcass during removal of the viscera. Birds should be caught and placed in crates or coops during the night to avoid excitement and possible injury prior to slaughter. However, they should be allowed fresh water until the last moment.

If you are not planning to do the deed yourself, look away now and continue reading about plucking. Many people don't want to slaughter chickens for obvious reasons and take their birds to a nearby farmer or poultry expert who will know exactly how to do it.

The bird should be killed in such a way as to allow most of the blood to drain from the body and at the same time limit struggling. A friend's father who keeps chickens stuns them in the back of the head with a broom handle before wringing their necks, but there are experts who say that wringing doesn't produce a properly bled carcass. A widely used method that accomplishes these objectives is making a cut just behind the jaw. The cut should sever the jugular vein without cutting the oesophagus or windpipe. A weight can then be hung from the beak to limit the movement of the bird. This is a more "humane" method of slaughter because the bird is unconscious due to loss of blood from the brain.

Some chicken keepers prefer the simple and straightforward "axe and chopping block" method. The block can be any chunk of firewood, as long as it's solid and squared off on both ends so that it won't move around or tip over and cause you to hurt yourself or maim a chicken as you wield your axe or hatchet.

Hold the tips of each bird's wings (to give you more control) right along with its feet in one hand as you position the chicken's head (with its neck well stretched out) on the block. One quick, firm, well-placed blow with a sharp axe or hatchet, then, is all you should need to sever the bird's head.

Continue to hold the chicken with its neck down so that it will bleed well. The blood can be allowed to fall around the bases of your fruit trees, if you have any, where it will both serve as a good fertilizer and discourage rabbits from killing the trees by nibbling away at their bark. If you need to, hold both the bird's wings and feet during this bleeding process to keep it from flopping around and spraying blood on you.

PLUCKING A CHICKEN

As soon as the chicken is bled—but before it has a chance to stiffen—you should carry it directly to a

waiting pile of old newspapers, a sharp paring knife, a bucket of boiling water and a pan of cold water for scalding, plucking, and washing.
Just remember that if your water is too cold and/or you do not immerse a chicken in it long enough, the bird will be hard to pluck. If the water is too hot and/or you leave your poultry in it too long, the feathers will practically fall out by themselves, but the skin of each plucked bird will be discolored and may even break in several places. Don't worry. The chicken will still be edible and practice makes perfect.

Grasp the body of the bird in one hand and one of its legs at the top of the thigh in the other.
Then pull the second hand down towards the foot. Most of the feathers will slip right off the leg in one operation. Repeat this process on the other leg, both wings, and the neck. If the chicken was scalded properly and didn't have too many pinfeathers, most of the body feathers can then be pulled out by the handful.

You are now ready to singe away the fine hairs that are still on the chicken. This can be done over an open gas flame. Or just twist some newspaper together, strike a match, and singe your birds outside if the wind isn't blowing too hard.

Remove the plucked and singed chicken's feet by bending each foot back and cutting through the joint. Some people like to skin the feet and cook them for soup or to add flavor to stock. Others just throw the unskinned feet to the dogs.

GUTTING A CHICKEN

The body cavity can be opened by making a small cut near the vent, extending the cut around the vent, being careful not to cut the intestine or contaminate the carcass with fecal material.

The abdominal opening should be as small as possible to improve the appearance of the finished product.

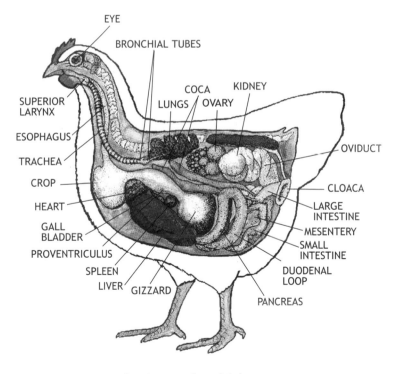

Anatomy of a chicken

After the abdomen is open, the viscera can be removed through the opening. It is very important to remove all the viscera, including the lungs, which are attached to the back. After all the contents of the cavity are removed, the bird should be thoroughly washed inside and out.

After the viscera have been removed, the heart, liver and gizzard should be separated and saved. The ends of any parts of the vascular system that may be attached to the heart should be removed by trimming off the top to expose the chambers. The heart should be washed and squeezed to force out any remaining blood. The green gall bladder should be carefully trimmed away from the liver. Next the gizzard should be split lengthwise and the contents washed away. The lining should then be peeled away from the rest of the gizzard.

After the gutting is finished, the carcass should be cooled as soon as possible. Iced water or a refrigerator can be used, however the iced water will do the job a little faster. If birds are to be frozen, the gizzard, heart, and liver can be wrapped in parchment paper and placed inside the body cavity. Many people also include a small packet of nutmeg to improve taste during storage. The birds can then be placed in a freezer bag and frozen.

USING YOUR EGGS

MUST EGGS BE REFRIGERATED?

Eggs sold in supermarkets in the United States are kept in chiller cabinets, and then stored in the refrigerator at home. Most people take them out of the carton and put them in the refrigerator door, which is usually fitted with an egg holder. However, some say that this is not where eggs should be stored. Now someone has come up with a system for laser dating eggs, since once you take them out of the carton, you have no idea how old they are. This is science gone mad! Surely if you are buying eggs, you buy only as many as you are going to use in a short period of time.

When I moved to England, I don't remember whether British refrigerators had egg holders or not, but shops that sold eggs (grocers and butchers) didn't refrigerate them. Nor did any of my neighbors. At first I stored my eggs in the refrigerator, but after my friends convinced me that it was just my American craziness to refrigerate everything, I stopped. I haven't refrigerated any eggs for over 20 years now, and I live to tell the tale, but commercial egg cartons clearly say that eggs should be refrigerated after purchase.

So what advice will I offer? If you know when your eggs were laid and they don't have any broken shells and you use them regularly enough so that they don't sit around for more than a couple of weeks, I say it is not necessary to refrigerate them. If you are at all worried, do refrigerate, in an egg carton in the main body of the refrigerator—but not in the refrigerator door!

STORING EGGS

FREEZING

There may be some discussion about this, but eggs can be frozen as long as it is done properly. There are three different methods of doing this.

You can freeze a whole egg: break the chosen number of eggs into a freezer container and scramble them with a fork (combine the yolk and white). Label the container with the amount of eggs and the date you are freezing and use within six months.

Freeze egg whites: separate the eggs carefully, weigh the whites, and place in a suitable container. Label with the weight and the date you are freezing, and use within six months. A typical large egg white weighs just under $1^{1}/_{2}$ ounces. If you are planning to use the whites in savory dishes, you can add a pinch of salt to keep the egg white from thickening.

Freeze egg yolks: separate the eggs, weigh the yolks, and place in a suitable container. Label with the weight and the date you are freezing, and use within six months. A typical large egg yolk weighs just under $3/4$ ounce. If you are planning to use the yolks in sweet dishes, you can add $1/2$ teaspoon sugar to keep the yolks from thickening.

PRESERVING

A tried and true method of preserving eggs is this: whatever excludes the air prevents the decay of the egg. What I have heard to be the most successful modern method of doing so is to place a small quantity of salted butter in the palm of the left hand and turn the egg around in it, so that every pore of the shell is closed; then dry a sufficient quantity of bran in an oven (be sure you have the bran well dried). Then pack the eggs in a box with the small ends down, in a layer of bran and another of eggs until your box is full. Shake to ensure that the bran fills all the spaces, cover the box, then place in a cold, dry place. If done when newly laid, the eggs will retain the sweet milk and curd of a newly laid egg for at least 8 or 10 months. Any oil will do, but salt butter never becomes rancid, and a very small quantity of butter will do a very large quantity of eggs.
To ensure freshness, rub them when gathered in from the nests; then pack when there is a sufficient quantity. The whole secret lies in carefully selecting fresh eggs, packing on end and keeping the air from them.

PICKLING

Pickled eggs are quite popular as a pub snack in the UK, with salad or cold dishes. Choose very fresh eggs and hard boil them. Cool under the cold water tap so that you can pickle them immediately. Remove the shells. You will find a recipe for pickled eggs on page 109. Place the eggs in an airtight jar with a lid and boil the vinegar and water with the spices; when the liquid is reduced and cooled, pour it over the eggs. Store in the refrigerator.

NUTRITIONAL VALUE

Eggs get a lot of bad press, but despite this, they are just about a "perfect" food containing almost all the nutrients necessary for life, lacking only vitamin C. Eggs are also a very inexpensive source of protein and nutritionally more efficient than cow's milk, fish, beef, or soya.

A large egg, weighing about a little over 2 ounces is approximately 30 percent yolk and 60 percent white, with the remainder being shell. It is roughly equivalent in protein to 1 ounce of lean meat, poultry, fish, or legumes.

The yolk contains proteins, fats, pigment, and minor nutrients. It also contains lecithin, which acts as an emulsifier for making mayonnaise and hollandaise sauce. Unfortunately, most of an egg's cholesterol is also found in the yolk, but there is now some controversy about whether it is "bad" cholesterol.

For almost 20 years, scientists have encouraged us to limit egg yolks in our diet because they contain cholesterol. Dietary cholesterol was thought to boost blood cholesterol levels, which increased the risk of heart disease. However, they are now re-evaluating this theory since many newer scientific studies point to saturated and trans fats as the major culprits behind heart disease.

An egg has very little fat, with most of it unsaturated and none of it trans fat. So most of us can again enjoy eggs without guilt.

Some people should limit their cholesterol consumption. These people usually have both high cholesterol and high triglyceride levels. They can still eat egg whites.

Egg whites are composed of water combined with several different kinds of protein. Egg protein is complete, which means it contains all the essential amino acids. It is perhaps the highest quality protein found in food, second only to mothers' milk.

To help prevent heart disease, it is more important to limit fats, particularly saturated fats and trans fats. So when you have eggs, eat them with grain foods, vegetables, and fruits, rather than more fatty fare.

USING YOUR EGGS

HOW TO BOIL A PERFECT EGG

This is a question that has vexed the minds of the greatest chefs in the world. A survey last year by the magazine *Waitrose Food Illustrated* asked five chefs how to boil an egg and received five different answers.

A high-profile story about soft-boiled eggs was in the media a couple of years ago. According to the broadcaster and author Jeremy Paxman, Prince Charles is very fond of a boiled egg and is very exacting about how he wants it cooked on any particular day. In his book, *On Royalty*, Paxman says Charles has his staff cook seven eggs ranging from runny to rock hard, and then tests them to see which one he wants. This information supposedly came from a friend of the Prince. However, Clarence House denied the claims, saying that the environmentally conscious Prince would never be so wasteful and that the story is untrue. This is not the first story about Charles's fondness for eggs. One courtier claims that the Prince's chefs cook several eggs in the morning so that whatever time he comes down to breakfast, there is an egg waiting for him. Another claim is that while he is out hunting, his staff will keep cooking batches of soft-boiled eggs so there will be one ready for him when he returns!

In my opinion, boiling an egg is not that straight-forward. Everyone has a different method — from the world's greatest chefs to the household cook.

I have never been all that good about soft-boiling eggs. For one thing, I don't like them cooked that way, so I've never learned. I think that starting the egg off in cold water is a good idea, because the shell is less likely to crack, but four minutes after the boil sounds like a long time, even for a very large egg.

I feel much more confident about hard-boiled eggs. My eggs are always room temperature. I poke a hole in the round end with a pin, put them into a pan in a single layer, cover with cold water, bring them just to the boil and barely simmer them for 10 minutes. This works perfectly for me with large eggs. Larger or smaller eggs will need some slight adjustment up or down.

When the time is up, take the pan and put it under the cold tap until the eggs are cool enough to handle, then remove them one by one and roll them around on a flat surface to crack the shells. Start peeling from the round end. You can thus remove the shells neatly from even very fresh eggs.

A bit of trivia about hard-boiled eggs: if you have cooked eggs, but are unsure about whether a particular egg is hard-boiled or raw because you might have put them together inadvertently, place the egg on its end and try to spin it like a top. The hard-boiled egg will spin and the raw egg will fall on its side and not spin.

OLD-FASHIONED PICKLED EGGS

You will need:

4–6 whole cloves

1 cinnamon stick

16 fluid ounces white vinegar

$1/2$ teaspoon mustard powder

$1/2$ teaspoon salt

$1/2$ teaspoon pepper

6 hard-boiled eggs, peeled

1. Put the cloves and cinnamon stick in a pan with the vinegar and bring the mixture to the boil.

2. Blend the mustard, salt, and pepper with a little water to make a paste and stir into the boiling vinegar. Simmer for 5 minutes.

3. Place the eggs in a jar with a wide mouth and a tight-fitting lid. Pour the hot vinegar mixture over to cover. Screw the lid on tightly and refrigerate for up to six months.

POACHING AN EGG

Simmering an egg in water will give you a firm white and a runny yolk—perfect to break over a salad and a necessary part of Eggs Benedict. As poaching them requires no fat, it's also a healthy way of eating eggs, although not if you're smothering them with Hollandaise sauce!

Here are three ways to do it:

Method 1
Bring a pan of water to a rolling boil and add a little white vinegar (1 teaspoon to $2^{1}/_{4}$ pints of water). This will help set the egg white. However, I don't like vinegary eggs much, so I don't add vinegar.

You want your eggs as fresh as possible, and at room temperature as this will mean they have firmer whites. Crack each egg into a cup, being careful not to break the yolk. Once the water is boiling, stir it to create a whirlpool effect and pour the egg in. The swirling water will wrap the white around the yolk. Simmer for about 5 minutes until the white is set, but the yolks are still runny.

Method 2
Follow the previous directions, but instead of simmering the egg, turn the heat off and leave to cook for 10 minutes.

Method 3

I like the idea of this technique the best and this is the one I use all the time now. Line a cup with a piece of microwaveable non-PVC plastic wrap large enough to wrap around the egg. Crack the egg into the lined cup and tie the plastic wrap around the egg.

Drop the egg into boiling water and simmer as usual. It comes out neat and perfect, and you won't have to scrub cooked-on egg off the pan.

HOW TO CODDLE EGGS

Coddled eggs are made by briefly immersing an egg in its shell in boiling water to slightly cook (or coddle) it. Coddling is similar to boiling, but instead of the egg being placed directly into the water, it is put into a lidded container called a coddler. The best eggs for coddling are the freshest eggs you can find. (If eggs are more than a week old, the whites thin out.) Whites of fresh eggs will gather compactly around the yolk, making a rounder, neater shape.

WHAT IS AN EGG CODDLER?

An egg coddler is a porcelain or pottery cup with a lid that is used to prepare coddled eggs. The egg(s) are broken into the buttered coddler, and seasonings are added. The coddler is then closed with the lid and partially immersed in boiling water for a few minutes. When the eggs are cooked to the desired firmness, the coddler is lifted from the boiling water, the lid removed, and breakfast is served in a lovely decorated dish.

How to Use an Egg Coddler

Bring a large pan of water to the boil. Butter the inside of the coddler and the inside of the metal lid. Break one or two eggs (according to size of the coddler) into the cup, and season to taste with salt and freshly ground pepper.

Screw on the lid (not too tightly—a loose turn is sufficient) and stand the coddler in the pan of boiling water, taking care that the water level comes almost all the way up the porcelain body of the coddler and stops just before the lid. Simmer for 7—8 minutes.

Remove the coddler from the water using the end of a fork/spoon through the lifting ring or lift using oven mitts or other heatproof gloves. Set the coddler on a towel or trivet. Using a towel and holding the lid by the rim, not by the lifting ring, twist the lid to loosen it. Serve the egg in the coddler.

FRYING AN EGG

I've had so many badly fried eggs in hotels an restaurants that I never order them anymore.
In fact, I rarely order eggs of any sort because I am so picky about them. This is how I fry eggs: I have a Le Creuset crêpe pan that I only cook eggs in, and I never allow anyone else to wash or dry it. It has never seen a scouring pad or anything that could scratch it.

If you're cooking with gas (which to my mind is the only way), heat the pan over a medium flame for a minute. Brush with butter until you see little bubbles, but before the butter burns.

Crack the eggs into the pan. My pan holds two comfortably. Make sure the heat is just high enough to cook the egg. When the egg white starts to solidify, slide a spatula designed for non-stick pans (wide enough to lift the egg) underneath. Do this carefully so you don't break the yolk. Take the pan off the heat and flip the egg to cook the other side very briefly and slide onto a plate, over toast or not, as you choose. In the USA this is called "over easy," and a fried egg cooked yolk-side is called "sunny side up." The former seems to suit the people I cook for, who like a runny yolk but not runny egg white.

SCRAMBLED EGGS

Never one to be overly modest, I tell people I make the best scrambled eggs in the world. In any event, I make really good scrambled eggs. It's very simple.

Heat a small non-stick pan over a medium flame for a minute. Add a small pat of butter and brush it around the bottom and sides of the pan. Meanwhile, crack two eggs into a cup and add a dollop of cream and a good grinding of pepper. Whisk with a fork and pour into the pan. With a wooden spoon, or a fork designed for non-stick pans, immediately start swirling the egg mixture around the pan and when the curds are almost set, remove from the heat and serve immediately.

Immediate is the operative word here. I wait until my husband only has a mouthful or two of his fruit left and then I start cooking the eggs. You can either season them before or after cooking, but the diner must wait for the eggs, not the other way around—otherwise they will be overcooked.

MAKING AN OMELET

Making an omelet is similar to making scrambled eggs, except that water, not cream, is recommended for the mixture. Water turns to steam, producing a light, airy omelet. Omelets, like scrambled eggs, cook very quickly, so always have your filling ingredients chopped, cooked, and

ready before you begin cooking the eggs.
A proper pan is important for successful omelet making. It is better to prepare several individual omelets, rather than one large one, because you will find that the individual ones will be lighter, fluffier and easier to handle.

An 8-inch crêpe pan with sloping sides is ideal.
A good-quality non-stick coating simplifies the process, but you can give an uncoated pan an almost non-stick surface by treating it with salt.
Heat the pan, then remove from the heat.
Add 1 teaspoon salt and scrub the inside of the pan thoroughly with paper towels. Tip the salt from the pan and repeat until the salt stays white. The salt acts as an abrasive, leaving a smooth surface.
Wipe the pan clean and re-treat as necessary.

To make the omelet, whisk two or three eggs with 2–3 teaspoons water and season with salt and freshly ground black pepper. Heat your pan over a medium flame for a minute. Add a pat of butter and brush around the bottom and sides of the pan.
Add the egg mixture and cook, loosening the edges of the omelet by running a knife around the edge, allowing the uncooked egg to slide underneath.
Cook gently until the omelet sets, then fill with your prepared ingredients and fold in half.
Slide the omelet out of the pan onto a plate and serve immediately.

BAKED EGGS

Sometimes called "Shirred Eggs," this is a method of cooking eggs in the oven without their shells. In France, this method of cooking is called *oeufs en cocotte*. For individual servings of baked eggs, use ramekins or individual soufflé dishes that just fit the eggs plus the flavoring, food, or liquid. Preheat the oven to 325°F/Gas Mark 3.

For each serving, lightly butter an individual ovenproof baking dish or ramekin.

Break one or two eggs into each dish. Season with salt and freshly ground pepper. Spoon 1 tablespoon milk or cream over the eggs. This helps prevent the eggs from drying out.

Bake for approximately 12—15 minutes, depending upon the size and amount of eggs. Check them after about 10 minutes. When done, the whites should be completely set and the yolks beginning to thicken, but not hard. Remove from the oven, garnish with chopped herbs and serve immediately.

COOKING EGGS IN THE MICROWAVE

To cook eggs successfully in the microwave, there are a few points that must be kept in mind: the most important of which is, eggs cooked in their shells will explode. Even eggs taken out of the shell may explode because rapid heating causes a build-up of steam. Always use a wooden toothpick or the tip of a knife to break the yolk membrane of an unbeaten egg before microwaving to allow the steam to escape.

Covering the cooking container with a lid or non-PVC microwave-safe plastic wrap encourages more even cooking, and if you forgot to pierce the yolks, will help to confine the explosion. Here are some ideas. All cooking times are based on a full power output of 600–700 watts. Higher or lower wattages will require some time adjustment.

FRIED EGGS
Crack an egg into a lightly greased microwaveable cup. Prick the yolk with a wooden toothpick or the tip of a knife. Cover the cup with non-PVC microwave-safe plastic wrap. Cook on 50 percent power just until the eggs are almost done, about 2–3 minutes. Let stand, covered, until the whites are completely set and the yolks begin to thicken.

HARD-BOILED EGGS
Separate the yolks and whites and put into two lightly greased microwaveable cups or bowls. Stir the yolks with a fork. Cover each cup with non-PVC microwave-safe plastic wrap. Cook separately on 50 percent power, stirring once or twice, allowing 20–30 seconds per yolk and 30 seconds–1 minute per white. Remove from the microwave when slightly underdone. Let stand, covered, about 2 minutes. Cool long enough to

handle comfortably, then chop, or chill in the
refrigerator until ready to chop.

POACHED EGGS

Pour 3 fluid ounces of water into a small, deep
microwaveable bowl. Crack two eggs into the bowl.
Gently prick the yolks with a wooden toothpick or
the tip of a knife and cover the bowl with non-PVC
microwave-safe plastic wrap. Cook on full power
for 1^1/$_2$–2 minutes. If necessary, allow to stand,
covered, until the whites are completely set and
the yolks begin to thicken, about 1–2 minutes.
Remove the eggs with a slotted spoon.

SCRAMBLED EGGS

Crack two eggs into a lightly greased microwaveable
cup or bowl. Whisk them together with
2 tablespoons of cream, and salt and freshly ground
pepper if desired. Cook on full power for 30
seconds–1 minute until they start to set, then stir
them and return to the microwave until almost set,
then remove and stir again. If necessary, cover with
non-PVC microwave-safe plastic wrap and allow to
stand until eggs are thickened and no visible liquid
remains, about 1 minute.

HOW TO SEPARATE EGGS

Set out two bowls. Crack the egg gently on a flat
surface or on the rim of a bowl, as close to the

middle of the egg as possible. If you crack it on the rim of a bowl it might be easier to get the egg to crack right in the center, but you are more likely to get pieces of eggshell in the egg whites. I crack eggs on a flat surface.

Working over a small bowl, use your thumbs to gently pry the egg halves apart. Let the yolk settle in the lower half of the eggshell, while the egg whites run off the sides of the egg into the bowl. Gently transfer the egg yolk back and forth between the eggshell halves, letting as much egg white as possible drip into the bowl below.
Be careful so as not to break the egg yolk.
Place the egg yolk in a separate bowl.

If you are planning to whisk the egg whites for a recipe, you might want to separate the eggs one by one into a smaller bowl, and then transfer the separated egg into larger bowls. This way, if you break a yolk it will not break into all the egg whites you've separated. The fat in the egg yolk (or any oil or grease) will interfere with the egg white's ability to whisk up properly. For this reason also you should wash your hands carefully to remove any natural body oils before working with egg whites.

If you get a piece of eggshell in the separated eggs, fish it out with a larger piece of shell.

Note that chilled eggs are easier to separate (the yolk doesn't break as easily), but most recipes call for working with eggs at room temperature.

So, you either let your eggs get to room temperature before separating them, in which case you'll need to be a bit more careful with the egg yolks, or let the eggs get to room temperature after you've separated them, in which case you should cover them in their bowl with plastic wrap and use them as soon as they get to room temperature.

EGG RECIPES

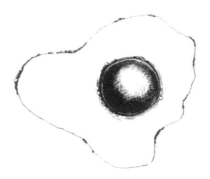

CAUTION

Some of these recipes use raw eggs. I am
assuming you are using eggs from your own
chickens. If not, pregnant women, the elderly,
very young, or those with a compromised
immune system, should avoid eating purchased
uncooked eggs.

MAYONNAISE

Mayonnaise is a simple emulsion of oil and egg yolks with a little acidity and salt added to brighten the flavors. Egg yolks contain lecithin, a natural emulsifier, which helps thicken sauces and binds ingredients. Homemade mayonnaise has a milder, more neutral flavor than store-bought and can be customized to your taste. You can either make mayonnaise by hand with a whisk and a bowl or in a food processor. The secret is to add the oil very slowly in a steady stream while the processor is running or whisking vigorously.

You will need: **Makes 8 ounces**

  1 egg

  Juice of 1 lemon *or*

  1–2 tablespoons vinegar to taste

  Pinch of salt and freshly ground black pepper

  8 fluid ounces light olive, sunflower, or corn oil

  Water to thin mayonnaise

Method:

1. Separate the egg and reserve the white for another recipe.
2. Combine the egg yolk and lemon juice or vinegar in a bowl, whisking to mix.
3. Continue to whisk constantly, adding the oil in a slow, steady stream.
4. If the mayonnaise looks too thick, add enough water, little by little, about 1 teaspoonful at a time, whisking after each addition.

5. When all the oil is added, the mayonnaise should look thick and fluffy, with the whisk forming ribbons through the mixture.
6. Store the mayonnaise covered in the refrigerator and use within five days.

N.B. If the mayonnaise doesn't thicken properly, you will have to start over or try to save the emulsion by adding another egg yolk and whisking vigorously again. You can adjust the seasoning with salt, pepper, and more lemon juice or vinegar if needed.

HOLLANDAISE SAUCE

Hollandaise sauce is the perfect accompaniment for Eggs Benedict, poached salmon, or steamed asparagus. The best way to make a perfect hollandaise is to do it in a blender. You can also use the same method with a double boiler.

You will need: **Makes about 6 ounces**

3 egg yolks

$^1/_2$ teaspoon Dijon mustard

1 tablespoon lemon juice

1 dash Tabasco or Worcestershire sauce

4 ounces butter

Method:
1. In a blender container, combine the egg yolks, mustard, lemon juice, and Tabasco or Worcestershire sauce.

2. Put the butter in a microwaveable cup and heat for about 1 minute on high, or until completely melted. Set the blender to high speed and pour the butter into the egg yolk mixture in a thin, steady stream. It should thicken almost immediately.
3. Keep the hollandaise warm until serving.

SCOTCH EGGS

Scotch eggs used to be part of a traditional Scottish breakfast, but most of them are now mass-produced and not worth eating. Homemade Scotch eggs are in a different league and are especially good eaten outdoors at picnics.

The basic recipe is sausage meat wrapped around a hard boiled egg, which is then coated in breadcrumbs and deep-fried. This is a dish that needs good sausage meat. It is probably best to take your favorite sausages and remove the casing.

Once you have mastered the basic recipe, it can be tweaked to produce a range of variations.

You will need: **Serves 4 as a snack**

5 eggs—the fresher the better, 4 for the scotch eggs
      and 1 for the coating
9 ounces pork sausage meat (if you are using sausages
      remove the casing by slitting them down one side,
      discard the skins)
Small onion, finely chopped

Butter or oil to fry onion

Spices and herbs—this will depend on the sausage meat you are using—(parsley, sage, or thyme and a bit of grated nutmeg, for example)

Breadcrumbs (either homemade or purchased)

Salt and freshly ground black pepper

Peanut or sunflower oil for deep-frying

Method:

1. Hard-boil four eggs. Cover them in cold water, bring to the boil and simmer gently for 8—10 minutes, depending on the size of the eggs. Drain and cool the eggs under cold running water. Set the eggs aside and allow to cool.

2. Fry the onion in butter or oil until softened and beginning to brown.

3. Combine the sausage meat, cooked onion, and herbs, and season with salt and pepper.

4. Peel the hard-boiled eggs. Beat the reserved raw egg. Divide the sausage meat into four portions, and pat each piece out on a floured surface to approximately 5 x 3 inches. Place a boiled egg in the center of each piece and carefully gather up the sausage meat to cover. Squeeze gently to seal. Coat the eggs by first dipping in the beaten egg and then rolling in the breadcrumbs. To make a crisper shell, you can repeat the dipping and rolling.

5. Heat $1^1/_2$ inches oil in a deep frying pan to a temperature of 350°F. (This is the temperature at which a small cube of bread will turn golden brown within a minute.)

6. Gently slide the eggs into the oil and fry for 6—8 minutes, basting and turning frequently until they have turned golden brown. Drain on paper towels to remove excess oil.

They can be eaten immediately, but are best served cold.

Variations:
1. Add finely chopped red bell peppers, chili powder, and paprika to the sausage meat.
2. Add small cubes of Cheddar cheese to the sausage meat.
3. Roll the dipped eggs in polenta or porridge oats instead of breadcrumbs.
4. Make a vegetarian version: instead of sausage meat, combine the following ingredients:

9 ounces grated Caerphilly cheese

5 ounces breadcrumbs

3 tablespoons finely chopped scallions

1 egg, plus 1 egg yolk

A heaped tablespoon of herbs (e.g., parsley, thyme, and sage)

1 teaspoon mustard powder

Chill in the refrigerator to firm up, then proceed as above.

DEVILED EGGS

You will need: **Makes 6 servings**

  6 eggs

  $1/2$ teaspoon paprika

  2 tablespoons mayonnaise

  $1/2$ teaspoon mustard powder

Method:

1. Place the eggs in a single layer in a pan of cold salted water. Bring to the boil and cook for about 10 minutes, until they are hard-boiled. Remove from the heat, put under a tap of cold running water, drain, and set aside to cool.

2. Cut eggs in half lengthwise. Scoop out the yolks and mash them in a bowl with the paprika, mayonnaise, and mustard powder. Spoon or pipe the mixture back into the egg whites and serve.

N.B. This is a basic recipe for deviled eggs. Feel free to add garlic, onion, horseradish, relish, or whatever you fancy, and substitute Dijon mustard for the mustard powder if you prefer. You can also sprinkle the paprika over the eggs once they are back in the whites if you prefer not to have orange yolks.

EGGS BENEDICT

You will need: **Makes 6 servings**

3 muffins, split and toasted

1 ounce softened butter

6 slices ham

6 poached eggs (recipe on page 110)

1 recipe hollandaise sauce (page 123)

chopped chives (optional)

Method:

1. Butter the muffins and top each one with a slice of ham and a poached egg.

2. Spoon over a portion of hollandaise sauce.

3. Garnish with chopped chives if desired and serve immediately.

CHEESE SOUFFLÉ

You will need: **Makes 4-6 servings**

2 ounces butter, plus extra for greasing

4 tablespoons flour

12 fluid ounces hot milk

6 eggs, separated

salt and freshly ground black pepper

$^1/_2$ teaspoon mustard powder

2 ounces freshly grated Parmesan cheese

2 ounces Cheddar or Gruyère cheese

Method:

1. Preheat the oven to 400°F/Gas Mark 6.

2. Butter a $3^1/_2$ pint soufflé dish.

3. Melt the remaining butter in a medium pan over a medium heat. Add the flour and reduce the heat a bit. Cook the flour until it is golden, about 3 minutes. Whisk in the milk gradually, then cook until the mixture thickens, whisking constantly. Remove from the heat.

4. Stir in the egg yolks, salt, pepper, and mustard powder. Add the cheese and stir until smooth.

5. In a clean bowl, whisk the egg whites until they hold soft peaks. Stir a bit of the egg white into the cheese mixture to lighten it, then very gently fold in the rest of the egg white.

6. Turn the mixture into the prepared soufflé dish and bake until risen and browned on top, about 30–40 minutes. Check for doneness with a thin skewer; if it is still wet, bake another 5 minutes or so; if it is barely moist, it is done. Serve immediately.

ONION QUICHE

If you like your onions sweet and cooked, this is a wonderful treat. You can use bought shortcrust pastry, but making it is not at all difficult and offers superior results.

You will need: **Makes 6—8 servings**

For the crust:
7 ounces all-purpose flour, plus some for dusting
$^1/_2$ teaspoon salt
5 ounces cold unsalted butter, cut into small cubes
2 egg yolks
about 3 tablespoons ice water

For the filling:
2 ounces unsalted butter
2 pounds mild onions, sliced thinly
salt and freshly ground black pepper
1 teaspoon fresh thyme
6 eggs at room temperature
16 fluid ounces light cream, heated until just warm

Method:
1. Make the crust. Mix the flour, salt, and butter in the container of a food processor. Process until the mixture looks like breadcrumbs, about 15 seconds. Add the egg yolks and process until mixed.
2. Pour in the water by the spoonful and process until the dough comes away from the sides of the container and forms a ball. Transfer the dough onto a

cloth dusted with a little flour and form the dough into a disk. Dust both sides of the disk with the flour, wrap the dough in parchment paper and refrigerate for about 30 minutes.

3. Roll out the dough to a diameter of about 2 inches larger than your quiche pan and fit firmly into the pan, fluting the edges. Refrigerate again while you make the filling.

4. Preheat the oven to 425°F/Gas Mark 7.

5. Prick the base of the crust all over with the tines of a fork. Line with parchment paper and ceramic baking beans. Bake 10 minutes. Remove the baking beans and parchment and bake for another 5 minutes. Reduce the oven temperature to 325°F/Gas Mark 3.

6. Make the filling. Melt the butter in a large, deep frying pan, add the onions, and season with salt and pepper. Turn the heat to medium-high and cook, stirring frequently, until the onions are very soft and golden brown, about 20 minutes. Stir in the thyme and remove from the heat.

7. Whisk the eggs with the cream and stir in the onions. Pour the onion filling into the partially baked crust and place the quiche in the center of the oven on a baking sheet.

8. Bake for 30—40 minutes until the filling is just about set. Remove from the oven and allow to rest for a few minutes. Serve hot or at room temperature.

ARBROATH SMOKIE AND CODDLED HIGHGROVE EGG

You will need: **Makes 4 servings**

  6 ounces Arbroath Smokie

  10 fluid ounces heavy cream

  5 fluid ounces milk

  4 medium/large organic eggs

  watercress leaves

  toasted brioche

Method:

1. Preheat the oven to 400°F/Gas Mark 6.
2. Gently poach the Arbroath Smokie in the cooking liquor ($1/3$ milk and $2/3$ cream).
3. Remove from the heat, cool and flake the fish off the bone, retaining the cooking liquor.
4. Place $1\frac{1}{2}$ ounces flaked fish per portion in a cocotte or ramekin and then half fill the ramekin with the cooking liquor. Crack a fresh egg into the ramekin and bake in the preheated oven for 8–10 minutes until the egg is cooked.
5. Serve with fresh watercress leaves and warm toasted brioche "soldiers."

Created by Liam Parr, Head Chef, Conservatory Restaurant, Calcott Manor, near Tetbury, Gloucester, England.

PERFECT MERINGUES

When you have a surplus of egg whites, a good thing to do with them is make meringues. Whether they are soft and fluffy as on an American lemon meringue pie or like a crisp base for pavlovas, snowy white meringues are a heavenly delight.

At its simplest, a meringue is made from egg whites and sugar. Sometimes salt and an acid, such as lemon juice or cream of tartar, are added to stabilize the egg foam.

FRENCH MERINGUE (SIMPLE MERINGUE)
1. Use a clean, dry bowl. It must be grease-free, because even the slightest amount of fat will ruin a meringue. Glass, ceramic, stainless steel, and copper bowls are all suitable, although some purists will tell you that the bowl must be copper. Do not use plastic bowls, because even though they may appear clean, they may still contain trace amounts of grease or oil.
2. Bring your egg whites up to room temperature. Pour the egg whites into the bowl and add a pinch of cream of tartar for each egg white. If you are using a copper bowl to make your meringue, do not add any acid, as it can react with the copper and discolor the egg whites.
3. Whisk with a hand-held electric whisk to medium soft peaks. Beat in 2 tablespoons of superfine sugar for each egg white. Continue to beat until the egg whites are glossy and hold firm peaks when the

whisk is lifted. Adding sugar early in the beating process ensures a firmer, finer-textured meringue.

Pasteurizing egg whites is not a health concern when preparing a meringue that is going to be baked for longer than 10 minutes in a moderate oven 350°F/Gas Mark 4, but for butter cream icing, baked Alaska or meringue toppings for pies, the egg white should reach at least 140°F for safety's sake.

ITALIAN MERINGUE

Italian meringue is made with a sugar syrup. Sugar and water are boiled to the soft ball stage, at 240°F, and carefully poured in a thin stream into a bowl of whisked egg whites. The mixture is then whisked until cool. When ready to use, Italian meringue will be very stiff, smooth and satiny. Because of the constant whisking, the bowl cools quickly and the egg whites may not reach pasteurisation temperature. You can use an instant read thermometer to check after the first minute or so of whisking. People with a compromised immune system should be careful not to consume undercooked egg whites.

SWISS MERINGUE

Swiss meringue is made by combining sugar and egg whites and heating them in a bowl over a pan filled with simmering water that doesn't quite reach the bottom of the bowl. Whisk the sugar and egg whites enough to break up the whites, but not enough so that they form a foam. The sugar will melt and act as a protective shield against coagulation of the egg whites. Heat and whisk constantly until the temperature of the whites reaches at least 140°F or hotter. Remove the bowl from the heat and whisk the egg whites with an electric mixer until they form stiff, glossy peaks.

FORGOTTEN COOKIES

These are small meringue cookies with a hint of chocolate and optional nuts.

You will need: **Makes 24 cookies**

2 egg whites

5 ounces superfine sugar

1 teaspoon vanilla extract

pinch of salt

6 ounces plain chocolate chips

4 ounces chopped walnuts or pecans (optional)

Method:

1. Preheat the oven to 350°F/Gas Mark 4. Cover two baking sheets with aluminum foil.

2. Whisk egg whites in a small bowl until foamy. Gradually whisk in the sugar until the meringue forms stiff peaks. Add vanilla extract and salt. Fold in the chocolate chips and nuts, if using.

3. Drop meringues by teaspoonfuls onto the prepared baking sheets. Place in the preheated oven. Turn the oven off, but do not open the oven door for at least 6 hours or preferably overnight.

ANGEL FOOD CAKE

This is a great way to use up excess egg whites and my friend Kathryn's favorite choice for her birthday cake. It has a wonderful texture and delicate flavor and is a terrific accompaniment for fresh fruit or ice cream.

You will need: **Makes 8—10 servings**
- 3$^{1}/_{2}$ oz sifted all-purpose flour
- $^{1}/_{2}$ ounce cornstarch
- 13 ounces superfine sugar
- 14 fluid ounces egg whites (from about 10—12 large eggs)
- $^{1}/_{2}$ teaspoon salt
- 1 teaspoon cream of tartar
- $^{1}/_{2}$ teaspoon vanilla extract

For the glaze:
- 4 ounces sifted confectioners' sugar
- 2 tablespoons lemon juice

Method:
1. Preheat the oven to 300°F/Gas Mark 2.
2. Sift the flour together with the cornstarch three times, then sift this mixture with 5 ounces of the sugar.
3. Whisk the egg whites in a large bowl with an electric mixer at medium-high speed until frothy. Add the salt and cream of tartar and whisk again until the whites just form soft peaks. Whisk in the remaining sugar a couple of tablespoons at a time.

Add the vanilla extract and whisk again until the whites hold soft peaks.

4. Sift one quarter of the flour mixture over the whites and fold in gently but thoroughly, then add the remaining flour in small amounts in the same way.

5. Spoon the mixture into an ungreased 10-inch tube pan and smooth the top. Tap on a counter top to eliminate any air bubbles.

6. Bake until the cake springs back when touched with a finger, and a wooden toothpick inserted near the center comes out clean, about 75 minutes.

7. Remove from the oven. If the pan has feet, invert on a rack. If not, invert over the neck of a long-necked bottle. Cool for 2 hours.

8. Make the glaze. Sift the confectioners' sugar into a bowl and mix with the lemon juice until smooth.

9. When you are ready to take the cake out of the pan, slowly run a long thin knife around the edges of the pan and the tube. Top with a rack and invert. Place a piece of baking parchment under the rack, pour the glaze over and leave to set.

10. Serve with fresh fruit or ice cream.

BAKED ALASKA IN A CUP

If you can whisk an egg white, you can make these classy little individual desserts. They are quick and delicious and you can make only as many as you need.

You will need: **Makes 4 servings**

chocolate chip cookies broken into chunks

ice cream flavor of your choice

2 large egg whites

$1/4$ teaspoon cream of tartar

3 tablespoons superfine sugar

$1/2$ teaspoon vanilla extract

Method:

1. Cover the base of four ovenproof and freezer-safe cups, glasses, or ramekins with cookie chunks to a depth of about $3/4$ inch.

2. Fill the cup with the ice cream of your choice. Use a spoon to press the ice cream tightly into the cup. Place the cups of ice cream in the freezer for at least 30 minutes to get rock hard.

3. Preheat the oven to 475°F/Gas Mark 9.

4. Whisk the egg whites and cream of tartar until soft peaks form. Drizzle in the sugar as you continue to whisk. Add the vanilla extract. Beat until stiff peaks form.

5. Pile the meringue on top of the frozen ice cream. Completely cover the ice cream so that the ice cream is insulated from the heat. Repeat with the other three cups.

6. Place the cups on a large baking sheet.
(The baking sheet will help deflect heat from the
bottoms of the cups.) Bake for 2—4 minutes or until
the meringue is golden. Serve immediately.

Adapted from *The Prepared Pantry*

CARAMEL DE CRÈME D'AMANDE

You will need: **Makes 6—8 servings**

14 ounces superfine sugar

2 fluid ounces water

24 fluid ounces whole milk

8 fluid ounces heavy cream

5 large eggs

5 large egg yolks

$1/4$ teaspoon salt

$3/4$ teaspoon almond extract

$1/2$ teaspoon vanilla extract

Method:

1. Preheat the oven to 325° F/Gas Mark 3. Set out
a 2-pound loaf pan.
2. Combine 6 ounces of the sugar with the water in
a heavy-based pan and cook over a medium heat,
stirring until the sugar is dissolved. Bring to the
boil and boil, brushing down the sides of the pan
with a pastry brush dipped in cold water until the
syrup is a pale gold color, about 5 minutes.
Continue to boil, swirling the pan until the syrup is
golden brown, about 2 minutes more, then
immediately pour the caramel into the loaf pan to

coat the bottom and ½ inch up the sides of the pan.
Be careful as the pan will get very hot. Set aside to
allow the caramel to harden.

3. In a clean pan, bring the milk and cream to the
bare simmer over a medium heat.

4. Meanwhile, whisk together the eggs, egg yolks,
salt, and remaining sugar in a large bowl. Add the
milk and cream mixture in a slow, steady stream,
whisking constantly, then add the almond and
vanilla extracts. Pour the custard through a fine
mesh sieve into the loaf pan and cover the pan
with a double layer of aluminum foil.

5. Put the pan in a larger pan and pour enough
boiling water into the larger pan to reach halfway
up the sides. Bake until a knife inserted into the
crème 1 inch from the edge of the pan comes out
clean but the center is still a bit wobbly, about
75 minutes. The crème will continue to set as
it cools.

6. Remove the loaf pan from the water bath,
remove the foil, and cool completely on a
wire rack.

7. Cover the pan with plastic wrap and refrigerate
for at least 3 hours or overnight. Run a thin knife
around the edge of the pan and invert the dessert
onto a serving plate.

8. Cut into slices to serve.

CHOCOLATE MOUSSE

You will need: **Makes 6 servings**

1 ounce unsalted butter

4 ounces 70% semisweet chocolate, chopped

3 eggs, separated

2 ounces superfine sugar

4 fluid ounces heavy cream

$\frac{1}{2}$ teaspoon vanilla extract

unsweetened cocoa powder, to decorate (optional)

Method:

1. Melt the butter and chocolate together in a glass bowl in the microwave. Whisk with a fork until smooth, then set aside to cool.

2. Whisk in the egg yolks and refrigerate the mixture.

3. Whisk the egg whites with half the sugar in another bowl until stiff peaks form. Set aside.

4. In yet another bowl, whisk the cream with the remaining sugar and vanilla extract until soft peaks form.

5. Stir a bit of the egg white mixture into the chocolate mixture to lighten it, then fold in the remaining whites gently but thoroughly. Fold in the cream and refrigerate until well chilled.

6. Serve cold, spooned into coffee cups or dessert dishes, dusted with sifted cocoa powder if you like.

EGGNOG

You will need: **Makes 4 servings**

3 eggs, separated

2 tablespoons superfine sugar

$1/2$ teaspoon vanilla extract

15 fluid ounces whole milk

10 fluid ounces light cream

4 fluide ounces light rum or brandy

freshly grated nutmeg

Method:

1. Beat the egg yolks with the sugar until well blended. Stir in the vanilla extract, milk, cream, and rum.

2. Whisk the egg whites until soft peaks form. Fold them thoroughly into the egg and milk mixture and refrigerate until cold.

3. Serve in coffee mugs or glasses, dusted with freshly grated nutmeg.

DYEING EASTER EGGS
THE NATURAL WAY

Give the Easter bunny a run for his money by dyeing Easter eggs the natural way, using common foods and spices. The longer you soak the eggs in the following liquids, the more intense the colors will be. Remember to start with hard-boiled eggs and refrigerate until ready to use, but remember this will only work on eggs with a white shell.

If desired, before dyeing the eggs, draw shapes, pictures, or inspiring words on them with wax crayons. The wax won't absorb the color so the designs will show through. Simply draw a design onto your eggs and then dye them as you would any other Easter egg.

Rubber bands are all you need to make tie-dyed eggs. Use a collection of different sized bands. Wrap them, one at a time, around the eggs. Make sure to leave some of the eggshell exposed so it can be dyed. Once the eggs are the color you like, remove them from the liquid and let them dry. When they have dried completely, remove the elastic bands.

HOW TO DYE THE EGGS

Wash the eggs in warm soapy water to remove any oily residue that may impede the color from adhering to the eggs. Let eggs cool before attempting to dye.

You need to use your own judgment about exactly how much of each dye to use. Except for spices, place a handful (or two or three handfuls) of a dyestuff in a pan.

Add tap water to come up at least one inch above the dyestuff. Bring the water just to the boil, then reduce the heat to low. Let simmer about 15 minutes or up to an hour until you like the color. Keep in mind that dyed eggs will not get as dark as the color in the pan. Remove the pan from the heat.

Pour mixture into a liquid measuring cup. Add 2—3 teaspoons) of white vinegar for each 8 fluid ounces of strained dye liquid. Pour the mixture into a bowl or jar deep enough to completely cover the eggs you want to dye. Use a slotted spoon to lower the eggs into the hot liquid. Leave the eggs in the water until you are happy with the color, several hours or overnight. The longer the egg soaks, the deeper the final color will be. If you plan to eat the eggs, be sure to leave the eggs in the refrigerator to soak.

When the eggs are dyed, lift them out with a slotted spoon. Dry on a rack or in an egg carton, which works nicely as a drying rack. Be careful to handle the eggs gently and minimally as some of the colors can easily be rubbed off before the egg has dried. For a textured appearance, dab the still wet egg with a sponge.

Eggs dyed with natural dyes have a matt finish and are not glossy. After they are dry, you can rub the eggs with cooking oil or mineral oil to give them a soft sheen.

Color	Dyestuff
Blue	canned blueberries, boiled red cabbage leaves, purple grape juice
Brown	strong coffee, boiled black walnut shells, black tea
Golden brown	dill seeds
Orange/brown	chili powder
Green	boiled spinach leaves
Yellow/green	boiled yellow apple peel
Orange	boiled onion skins, paprika, carrots
Pink	cranberry juice, raspberries, liquid from pickled beets
Red	pomegranate juice, canned cherries
Violet/purple	red wine
Yellow	boiled lemon rind, chamomile tea, ground turmeric

CHICKEN
RECIPES

BASIC CHICKEN STOCK

Chicken stock is a great way to use the bones that are left over when you bone chicken breasts. You can also use a leftover cooked chicken carcass instead of fresh raw meat to make stock. Once you make chicken stock, making chicken soup or anything requiring a chicken-flavored sauce is a doddle.
As for the vegetables that go into chicken stock, you can use whole fresh ones, or save leftover scraps. If you want a darker, richer stock, roast your chicken, chicken bones and vegetables in a 450°F/Gas Mark 8 oven for about 40 minutes before adding them to your stockpot.

So, here's the basic chicken stock recipe (you can use an approximate amount of scraps instead of the whole vegetables listed):

You will need: **Makes about 7 pints**
  4—5 pounds chicken portions or meaty bones
  1 large chopped onion
  2 or 3 large chopped carrots
  3 or 4 sticks celery (the leafy top parts are great for
      stock as well)
  6—8 chopped garlic cloves
  1 tablespoon whole black peppercorns
  10 pints cold water

1. Put all your ingredients into a pot and simmer for about 2—3 hours. Periodically skim off the foam as it rises to the top of your pot.

2. When finished cooking, strain the stock and refrigerate for a few hours. Any fat in the stock will congeal at the top and can be easily strained off. Your stock is now ready for use in a recipe or for the freezer.

CLASSIC CHICKEN NOODLE SOUP

Classic chicken noodle soup is a comfort food suitable for any time of the year, but especially during the cold winter months and when you are ill. It is often jokingly called Jewish penicillin. This recipe is made from scratch, so allow sufficient time to cook the chicken. It makes a lot of soup!

You will need: **Makes 12—14 servings**

2 tablespoons corn oil

2 medium onions, chopped

3 medium carrots, cut into rounds

3 celery sticks, chopped

1 fat hen, about 6—7 pounds, or 2 smaller chickens

$3^{1}/_{4}$ pints good-quality chicken stock

$1^{3}/_{4}$ pints cold water or more if necessary

4 sprigs fresh parsley

3 sprigs fresh thyme or $^{1}/_{2}$ teaspoon dried thyme

1 bay leaf

salt and freshly ground black pepper

8 ounces uncooked egg noodles

Chopped fresh parsley to garnish

Method:

1. Heat the oil in a large stockpot over a medium heat. Add the onions, carrots, and celery and cook, stirring often, until softened, about 10 minutes.

2. Cut the chicken into 8 pieces. Add the chicken to the pot and pour in the stock. Add enough cold water to cover the ingredients by 2 inches. Bring to the boil over a high heat, skimming off the foam that rises to the surface. Add the parsley, thyme, and bay leaf.

3. Reduce the heat to low and simmer, uncovered, until the chicken is very tender, about 2 hours.

4. Remove the chicken from the pot and set aside until cool enough to handle. Remove and discard the parsley, thyme, and bay leaf. Let stand 5 minutes and skim the fat from the soup.

5. Remove and discard the chicken skin and bones and cut the meat into bite-size pieces. Add the noodles and cook until done, about 10 minutes. Stir the meat back into the soup and season to taste with salt and pepper. Serve hot. (The soup can be prepared up to 3 days ahead, cooled, covered, and refrigerated, or frozen for up to 3 months.)

POACHED CHICKEN

My favorite sandwich is chicken and green salad leaves with a bit of mayonnaise on granary toast. This is how I prepare the chicken. It also makes a gorgeous chicken salad.

You will need: **Makes 2 servings**

2 boneless chicken breasts with skin

4 fluid ounces chicken stock

4 fluid ounces dry white wine

1 teaspoon Italian seasoning

black pepper

pinch of paprika

Method:

1. Preheat the oven to 400°F/Gas Mark 6.
2. Place the chicken breasts skin side up in a shallow baking dish and pour in the chicken stock and white wine. Sprinkle with Italian seasoning, grind some pepper over and dust with paprika.
3. Poach the chicken in the oven for about 30 minutes until the juices run clear when pierced with the tip of a knife. Remove from the oven and cool.
4. Remove the skin and slice or cube the chicken as needed.

SIMPLE ROAST CHICKEN

I rarely roast a whole chicken anymore, preferring to portion it before cooking because I'm lazy, but if I were going to, I would use this method, which I originally found in a Simon Hopkinson cookbook.

You will need: **Makes 4 servings**

4 ounces unsalted butter, softened

4 pounds free-range chicken

salt and freshly ground black pepper

1 juicy unwaxed lemon

Several sprigs of fresh thyme or tarragon, or a mixture

1 clove garlic, peeled and smashed

white wine for the sauce (optional)

Method:
1. Preheat the oven to 450°F/Gas Mark 8.
2. Smear the softened butter all over the skin of the chicken with your hands or a pastry brush. Put the chicken in a roasting pan and season with salt and pepper. Squeeze over the juice from the lemon. Put the herbs, garlic, and lemon rinds inside the cavity of the chicken. This will give it a delicious flavor.
3. Roast the chicken, uncovered for 15 minutes. Baste, then reduce the heat to 375°F/Gas Mark 5 and roast for an additional 45 minutes with occasional basting. The chicken will be golden-brown all over with buttery lemony juices in the roasting pan. Check to see that it is cooked through by piercing the top of the thigh with a sharp knife.

The juices should run clear.

4. Turn the oven off, leave the door ajar and let the chicken rest in the pan for at least 15 minutes. This will relax the flesh and make it easier to carve.

5. Pour the juices into a pitcher and stir to amalgamate. Carve or portion the chicken and serve with the buttery juices. There is no need to make gravy, but you could add a bit of white wine to the lemon butter sauce if you like.

CHICKEN CURRY

There isn't anything standard about an authentic Indian curry. Each cook in the locations in which curry is indigenous uses a different spice mix. This is a reasonably authentic Murg Kari, a southern Indian chicken curry. Feel free to adjust the quantity of spices to your taste.

Taste differs quite a bit. When I was in northern India, I happened to be part of a business lunch group in a local village restaurant. After looking at the menu, the man seated next to me and I decided we would share one of the curries that was cooked for two. When I enquired whether to order it mild, medium or hot, telling him I had a good tolerance for heat, he said, "Medium-hot will be much too spicy." When our curry arrived, the poor man kept wiping his face with his napkin, muttering, "Too hot! too hot!" I thought it was very mild.

You will need: **Serves 4—6**

About 2¼ pounds skinned chicken legs and thighs and
     boned and skinned chicken breasts cut in half

salt

2 ounces ghee or clarified butter

2 medium onions, chopped

1 clove garlic, finely chopped

2 teaspoons finely chopped fresh ginger root

1 teaspoon ground cumin

1 teaspoon turmeric

1 teaspoon ground coriander

1 teaspoon chili powder

½ teaspoon crushed fenugreek seed

4 fluid ounces water

2 ripe tomatoes, peeled, seeded and chopped

2 tablespoons finely chopped fresh cilantro

4 ounces plain yogurt

1 teaspoon homemade or purchased garam masala

1 tablespoon fresh lemon juice

Method:

1. Pat the chicken pieces dry with paper towels and sprinkle with salt.

2. Melt the ghee or clarified butter in a heavy frying pan over a medium-high heat, add the chicken pieces and fry for 3—4 minutes until white on all sides and fairly firm. Transfer the chicken to a plate using a slotted spoon.

3. Add the onions, garlic, and ginger to the pan and cook for 7 or 8 minutes, stirring constantly until the onions are soft and golden brown. Reduce the heat to low and add the cumin, turmeric, ground coriander,

chili powder, fenugreek, and 1 tablespoon of the water and fry, stirring for a minute or two. Stir in the tomatoes, half the fresh cilantro and the yogurt. Season with salt if necessary.

4. Increase the heat to medium and add the chicken and any juices that have accumulated on the plate. Pour in the remaining water. Bring to the boil, turning the chicken in the sauce to coat evenly. Sprinkle with garam masala and the remaining fresh cilantro, reduce the heat to low, cover tightly and simmer for 20–25 minutes until the chicken is very tender. Sprinkle with the lemon juice and taste for seasoning.

5. Serve with fluffy white rice. My children also liked the little bowls of accompaniments I used to serve with curry, e.g., peanuts or cashew nuts, mango chutney, raisins, flaked coconut, sliced bananas and chopped hard-boiled egg.

CHICKEN POT PIE

You will need: **Makes 4—6 servings**

1½pounds boneless, skinless chicken breasts or thighs

salt and freshly ground black pepper

1 pint chicken stock

1 tablespoon olive oil

12 small white onions

bouquet garni

juice of 1 lemon

24 white mushrooms

8 ounces fresh or frozen peas

1 ounce butter

4 tablespoons flour

10 fluid ounces heavy cream

3 tablespoons dry white wine or sherry (optional)

9 ounces unsweetened pastry

1 large egg yolk

Method:

1. Season the chicken with salt and pepper and put into a casserole with the stock. Bring to the boil over a medium heat and simmer, covered, until the chicken is cooked, about 10 minutes. Remove from the heat. Strain the stock into a bowl and transfer the chicken to a large bowl.

2. Increase the heat to medium high and add the olive oil to the casserole, then add the onions, bouquet garni, lemon juice, mushrooms, and peas. Sauté until just tender. Meanwhile, cut the chicken into bite-size pieces. Transfer the vegetables to the bowl with the chicken. Discard the bouquet garni.

3. Preheat the oven to 400°F/Gas Mark 6.
Make the sauce. Melt the butter in the casserole over a medium heat, add the flour and cook for 1 minute. Whisk in the reserved chicken stock, cream, and any accumulated chicken juices. Bring to the simmer, reduce heat to low and continue to simmer until the sauce thickens, about 1 minute. Taste for seasoning and stir in the wine or sherry if using.

4. Transfer the chicken, vegetables, and sauce to a pie dish.

5. Roll out the pastry slightly larger than the dish. Moisten the edge of the dish with a little beaten egg yolk and place the pastry over the dish, cutting off any excess pastry and fluting the edges. Make decorations with any scraps and stick to the pastry, brushing with egg yolk to glaze. Cut vents in the pastry to allow the steam to escape, and bake in the oven for 20–30 minutes, until the pastry is golden and the filling is piping hot.

CHICKEN LANGUAGE

Don't put all your eggs in one basket
Don't venture all you have in one speculation;
don't put all your property in one bank.

**Don't count your chickens before they
are hatched**
Don't anticipate your profits before they materialize.
One of Aesop's fables describes a market woman
saying she would get so much money for her eggs.
With this money she would buy a goose. This would
bring her so much, she would be able to buy a cow,
and so on. But in her excitement, she kicked over her
basket of eggs and broke all of them.

She's no spring chicken!
She's no longer young. The implication is that she has
reached an age when she is no longer a chick. This
appeared in *The Spectator* in 1711 which was quoted
as, "You ought to consider you are now past a chicken;
this Humour, which was well enough in a Girl, is
insufferable in one of your Motherly Character."

Egg on
To encourage someone and urge them forward.
This is a variant of "edge," to egg someone on is to
edge them forward.

Teach your grandmother to suck eggs
Trying to offer advice to those who are older and
have more experience. The French expression is

"Les oisons veulent mener les ois paître" (the goslings want to drive the geese to pasture).

Walking on eggshells
Being very cautious as if you were walking on eggshells, which are easily broken.

Egg on your face
Made to look foolish or be embarrassed; to do something ineptly. The expression originated in the USA, probably due to the fact that someone eating an egg sloppily is likely to end up with some of it on his face.

Pecking order
Pecking order among hens has a definite prestige pattern. Hens, like humans, freely peck at other hens below their rank and submit to pecking by hens above them. Hens rarely peck at roosters in the barnyard, where they are the "cock of the walk," but in the 17th century, it was believed that hens often pulled feathers from roosters below them in the pecking order.

Playing chicken/chicken out
Imagine two fast cars on a deserted stretch of road. At a signal, they both drive at high speed towards each other down the middle of the road. The one who "chickens out," i.e., turns away, sometimes with the consequence of driving into a ditch, loses.

A cock and bull story

A fanciful and unbelievable tale. It is said that the phrase originated at Stony Stratford, Buckinghamshire, UK. Coaches between London and Birmingham used to change horses in the town at two of the main coaching inns—the Cock and the Bull. The banter of the rival groups of travelers apparently resulted in exaggerated and fanciful stories. While this explanation is plausible, there is no real evidence to support this claim. More likely it is from Robert Burton's *The Anatomy of Melancholy*, 1621: "Some men's whole delight is to talk of a Cock and Bull over a pot."

Cock a snook

A derisive, nose-thumbing gesture showing someone contempt or opposition. Nobody is entirely sure what a "snook" is in this context. It is sometimes reported to be derived from "snout," but then why is this expression not "cock a snout?"

A bad egg

Someone or something that disappoints expectations.

Nest egg

A special sum of money saved or invested for a specific purpose. Nest eggs are usually intended for retirement, education, and entertainment.
The main idea is that the money in the nest egg shouldn't be touched except for the purpose for which you saved it.